NATURE'S
DESIGN

exploring the mysteries
of the natural world

Richard Thompson

DEDICATION
This book is for my children – Julian, Sarah, Adrian, Katherine and Hannah
with much love. And for Deborah Carter, who changed everything.

ACKNOWLEDGEMENTS
This book owes its origins to many conversations with Colin Walker –
children's author, educationalist and friend in Wellington. Thanks to the reference
librarians at the Wellington Clinical School of Medicine for their diligent searches for often
obscure material. A large thank you to Helen de Villiers whose editorial skills created
order from chaos. And many thanks to Linda Crerar who typed the original manuscript.

Struik Nature
(an imprint of Random House Struik (Pty) Ltd)
Cornelis Struik House
80 McKenzie Street
Cape Town
8001 South Africa

Company Reg. No 1966/003153/07

Visit us at **www.randomstruik.co.za**

First published in 2008

Publishing manager: Pippa Parker
Editor: Helen de Villiers
Designer: Janice Evans
Illustrator: Lucy Owen
Proofreader: Jane Smith
Indexer: Cora Ovens

Reproduction by Hirt & Carter Cape (Pty) Ltd
Printed and bound by CTP Book Printers

ISBN 978 1 77007 724 9

CONTENTS

CHAPTER ONE

EMERGING FROM THE FORESTS
the story of grass

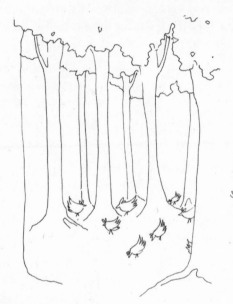

The chickens have, in the absence of their staple diet of grain, taken to the woods in search of alternatives.

A world without grass

Ours is a world of grass – think of our suburban lawns, our sports fields and inner city parks, our countryside with patchwork quilts of well-tended fields and windswept uplands dotted with sheep. On a much larger scale, give some thought to the great natural grasslands of the world – the prairies and pampas of the Americas, the steppes of Eurasia, the plains and savannahs of Africa and the Canterbury plains of New Zealand. Consider, too, that all the major cereal crops are members of the grass family, a list including wheat, barley, rye, oats, sorghum, maize, sugar cane and even rice.

With global temperatures predicted to increase by a matter of three to five degrees centigrade in this century, our grandchildren may have to confront a world where drought and heat have shrivelled the grasslands to scrub and desert. In this brave new world, a farmer draws the curtains to greet a new day. The view from his window is no longer one of fields full of lush meadow grass or crops bending in the wind, but the bleak aspect of barren earth spotted with weeds and other invaders. Of course, he may have planted acres of root vegetables and the fare of market gardens but, if he has not, then the land will slowly but surely return to the forests that flourished here long ago. In the farmyard he will find silence – the chickens, which in days past fussed and squabbled around his feet have, in the absence of their staple diet of grain, taken to the woods in search of alternatives. They are unlikely to survive the first winter. In days gone by, the farmer was able to send grain to the miller to be ground into flour, but the mill wheels are now immobile – there is to be no more bread. The barn will soon house a dusty museum of rusting and obsolete machinery. And there will be no more bales of hay prudently laid down at harvest time to feed the stock over a harsh winter; not that this matters particularly, because there is no longer any stock to feed. All the animals that depended on grass as a primary or sole food source are now dead: there are no horses, no cattle, no sheep and no goats. The only domesticated or semi-domesticated animals likely to survive are the pig and the forest deer.

Stretching in a broad sweep from Canada to Kansas, the great heartland of America is now a dust bowl, a place where only hardy plants such as the inedible sage and tumbleweed can survive and where trees are limited to the banks of permanent water courses. In the wake of these events the dairy industry is in ruins and a similar fate has befallen the meat industry and the woollen mills. The economies of the Americas and Europe are in tatters. Farmers in Asia, now faced with vistas of empty rice paddies, will dread the consequences of losing a crop that feeds over half the world's population. African farmers, too, having lost their maize, cattle and goats, which form the dietary cornerstone of a whole continent, will face starvation.

In the aftermath of such a global catastrophe, the fate of the world's wilder places would be sealed. The plains of Africa would be silent, the migrating herds consigned to memory – no flicker of tension rippling along the flank of a zebra before it explodes into full gallop, no growl from a tawny shape hidden in the undergrowth and no telltale dust cloud to give away the distant passage of a herd of wildebeest or buffalo. No sign, in fact, of any grazing antelope or gazelle, no trace of zebra, wildebeest, buffalo or hippopotamus. And in their wake would follow the predators that stalk them – lion, leopard, cheetah, hyaena and wild dog.

And, on a more personal note, there would be a large question mark hovering over the whole evolution of our own species. Would we even have existed in a world without grass? Ancestral man moved out from the tree line to forage on the grasslands and savannah of Africa, searching for fruits and roots and acquiring, in time, a taste for the flesh of other creatures. Initially hunting in groups to drive large predators from their kills, he gradually developed the necessary weapons and skills to become a hunter in his own right. Later, the necessity of a nomadic existence in the continual search for food gave way to a more stable society when primitive man acquired the ability to domesticate wild animals and to cultivate and harvest the seeds of wild grasses – a less risky and more reliable method of food production. But, in the absence of the grasslands, and hence of a readily available four-footed food supply or of seeds to sow, would he have ventured beyond the forest? What would have been the incentive? Confined to forest and woodland by availability of food, natural selection would not have favoured the emergence of a species with a straight spine and with hip joints enabling upright walking, which would have reduced climbing ability. Is it possible that our own evolution would have halted at this point, making no progress beyond that of the modern-day wild primates living on a diet of leaves, fruits, nuts, seeds and small monkeys?

Throughout the animal kingdom one organism must eat another for its very existence. This arrangement creates a thread running through the natural world that links the fate of one form of life inextricably to another. Nothing in nature occurs in isolation and the lives of plants, insects, birds and animals are interwoven in such a way that the success of any one species is largely dependent on the continued existence and well-being of others. Out on the African plains, the grass is eaten by the wildebeest, the wildebeest by the lion, the lion in its dotage may become the victim of the hyaena and the corpses of all of them are picked apart by vultures and small creatures. In a more domestic scene the cattle, sheep and goats that crop the grasses form the basis of a huge industry supplying dairy produce, meat and the raw materials for clothing. In both these instances grass is the focal point of a living system, the centre of a cycle of dependency that relies on the integrity of the whole for its well-being. One break in the sequence has obvious consequences further along the chain and, the earlier this break occurs, the greater will be the disruption.

*The grass is eaten by the
wildebeest, the wildebeest by
the lion ... and the corpses of
all of them are picked apart by
vultures and small creatures.*

The grasslands play a critically important role in the lives of many species, including our own. But where did they come from and how have the grasses enjoyed such success, managing to survive and flourish in the face of continual assault by grazing animals and frequently hostile climatic conditions?

All the grasses are flowering plants (angiosperms) belonging to the family Poaceae. Currently there are approximately 10 000 species of grass, which occupy a staggering 25 to 30 per cent of the terrestrial land surface. The original ancestor of the grasses was probably a flowering plant of the lily family (Liliaceae) living on the edge of the forests. Grasses do not flourish in the shade of dense woodland or forest, preferring instead drier open spaces with good light and wind – the necessary requirements for rapid growth and seed dispersal. Their entrance onto the world stage was therefore delayed until certain geological and climatic events opened a niche for them. The worldwide distribution of grasses has led to the view that they must already have been established prior to the break-up of the 'supercontinent' Pangea some 230 million years ago, but the first unequivocal fossil grass pollens have been dated to a much later period between 65 and 55 million years ago.

Mass extinction and the plant world

Curiously enough, the grasses emerged at, or very close to, the time of a major catastrophe for life on Earth, which resulted in the extinction of 60 to 75 per cent of animal species and included 90 per cent of all types of protozoans and algae, 80 per cent of marine invertebrates, half of the mammals and all of the dinosaurs. With few exceptions, the crocodile being one of them, all animals weighing more than 25 kilograms were rendered extinct. This event occurred 65 Ma (Mega-annum, or million years ago) at the boundary of the Cretaceous and Tertiary periods and became known as the K-T extinction (the K is derived from the German word *Kreidezeit*, meaning 'time of chalk'; the T is from 'Tertiary').

Extinction of species has been part of Earth's history from its beginnings, where one or more groups disappear as a result of more or less localised changes in climate, geography and food resources. These background extinctions need to be distinguished from the more dramatic mass extinctions, which involve large numbers and different types of species on a more global scale. The geological record points to as many as 15 such events in the past, of which five stand out head and shoulders above the rest. The K-T extinction has been the most studied of these, largely because of interest in the final demise of the dinosaurs.

As it happens, the Earth was undergoing major changes at the time of the K-T event. It is known that the Earth went through a period of intense cooling at this time and that sea levels fell by approximately 100 metres, creating land bridges and exposing continental shelves. This probably accounted for the death of many marine species that could only survive in shallow water. The climate was further altered by tectonic surges in the process of building mountain ranges such as the Rockies, Andes and European Alps, and some scientists have proposed that tectonic activity pushing Australia away from Antarctica resulted in the production of a deep, cold Antarctic current moving northwards towards the equator. This idea fits well with the observed pattern of extinction that predominantly affected tropical fauna, leaving many species in high latitudes intact. There have been numerous theories presented to account for the K-T extinction, but two of these seem to have captured the public imagination, as well as polarising the scientific community. They offer the alternatives of a massive asteroid collision with the Earth or a prolonged period of intense volcanic activity.

In the 1970s Luis Alvarez, the Nobel Prize winner, and his colleagues were studying the rock sequences of deep-water limestones at Gubbio in Italy. Here the K-T boundary is represented by a thin layer of clay measuring only one centimetre in thickness. Their intention was to determine whether the K-T event had occurred in a short space of time or had gradually unfolded over tens or hundreds of thousands of years. In this particular study, they were unsuccessful, but the clay layer did reveal some unexpected findings. Much to their surprise, chemical analysis showed a high concentration of the element iridium, a substance rare in the Earth's crust (0.3 parts per billion) but relatively common in the Earth's core and in meteorites (500 parts per billion). Examination of the adjacent rock layers showed no such increase in the levels of iridium, and the team were left to ponder the reasons for such an anomaly. Since the K-T boundary layer was so thin in places, being no more than one centimetre at times, it seemed unlikely that the iridium had been deposited over a prolonged period.

After considering a number of possible explanations, they were drawn to the conclusion that the source of the iridium could only have been the result of an asteroid or comet impact throwing huge quantities of iridium-rich dust into the atmosphere; this dust eventually would have settled all round the globe. Support for this idea subsequently came from the discovery of over a hundred other sites around the globe showing the same iridium anomaly at the K-T boundary layers. Particularly striking were the sharp iridium peaks found at Caravaca in Spain and in the Raton Basin in New Mexico. An extraterrestrial origin was also suggested by further chemical analysis, which demonstrated that the concentration ratios of iridium to other trace elements such as platinum and gold were typical of those found in most primitive meteorites.

The clay deposits in Spain and Italy also revealed two further pieces of the puzzle, the discovery of minute (one millimetre diameter) glassy mineral spheres and shocked grains of quartz. Mineral spheres such as this are known to be formed by the melting of rock following high-velocity, high-temperature impact, which ejects the melted particles up into the atmosphere. Cooling and condensation into glassy spheres occurs high in the atmosphere before the particles return to Earth. Shocked quartz with a characteristic laminated structure is produced typically by high-velocity, high-energy shock, and is found in impact craters and nuclear test sites.

Chemical analysis has suggested that mass extinction of the dinosaurs could have resulted from an extraterrestrial event.

From various calculations, Alvarez believed that an asteroid measuring approximately 10 kilometres in diameter would be necessary to account for the K-T extinction. Assuming it struck the Earth at 10 kilometres per second, its impact would have released energy equivalent to the explosion of a 100 000-megaton nuclear bomb, creating an impact crater almost 150 kilometres in diameter. Everything within visual range of the fireball would have perished instantly. Computer modelling of a collision of this magnitude shows that millions of tons of dust and debris would be thrown into the atmosphere, sufficient to cut out light and heat from the Sun. Restricted initially to one hemisphere or the other, depending on the site of impact, this would be a global phenomenon within 12 months (based on the rate of atmospheric spread of radioactive carbon, C^{13}, following experimental atmospheric nuclear explosions). An absence of light and heat would trigger an 'impact winter' characterised by intense cold and the cessation of photosynthesis. In the resulting cold and darkness, food chains could be expected to collapse and all large animals, particularly reptiles incapable of generating their own body heat, would quickly die out. Evidence from the fossil records shows that, in fact, most animals weighing 25 kilograms or more perished in the K-T extinction. A period of cold would have been followed by a period of global warming as a result of greenhouse gasses produced by forest fires, and an increase in atmospheric water vapour. Analysis of soot levels at the K-T boundary suggests the likelihood of extensive forest fires, equivalent to half the world's current forests. And if all this were not enough, the impact would have generated large tidal waves (tsunamis) and probably numerous earthquakes. The high temperature impact could also have led to the chemical combination of atmospheric nitrogen and oxygen forming oxides of nitrogen, which, when mixed with water, produces nitric acid rain. Such rainfall would have increased the acidity of surface oceanic waters, resulting in the wholesale death of plankton, a vital starting point in the food chain, the loss of which could only spell catastrophe for large numbers of marine species.

Needless to say, the consequences of such a catastrophic impact would have been felt immediately, with ongoing effects stretching out to hundreds of thousands of years. But does the available evidence for such an occurrence really stand up? Where is the impact crater and has a similar event occurred at other times in the Earth's history? Unfortunately, even with the most sophisticated modern radio isotope methods, the precise timing of past geological events is difficult: it is rare to give an accurate estimate of a given event to within 100 000 years and often the time range is much longer. It seems the K-T extinction can be timed to within 10 000 to 100 000 years simply because it occurred during a period of reversal of the Earth's magnetic field, but more precise measurements are not thought to be possible at the present time.

In 1991 a meteor impact site dating to the time of the K-T event was discovered in the region of the Yukatan peninsula in Central America. The impact crater was 180 kilometres in diameter, and analysis of core samples obtained by drilling in this area showed the presence of shocked quartz. Further drilling in the nearby Caribbean Sea off Haiti also led to the discovery of a layer of glass one metre in thickness. The glass was rich in sulphur and completely unlike any previously found in connection with known volcanic activity. A more likely explanation suggests that this glass originated from melting of the continental crust and was thus more typical of an impact site. Geological evidence of tidal waves has also been found in the southern USA adjacent to the impact area.

The Earth has witnessed two well-documented large meteor collisions in the past, neither of which was associated with a mass extinction event or an iridium anomaly. One of these occurred 230 million years ago (Ma) and led to the formation of the 70-kilometre-wide Manicouagan crater in Canada, while the more recent collision occurred in Popigai, Siberia, some 40 Ma, leaving a huge crater 100 kilometres in diameter. There is therefore some room for doubt concerning an extraterrestrial cause for the K-T event. Can the same be said for a volcanic origin?

In northwest India lies an area known as the Deccan Traps, which is made up of layered flows of basaltic lava that may, at one time, have covered an area of two million square kilometres. In places, the laval sediment is almost two and a half kilometres in thickness and the area it now covers is roughly half the size of Europe. Research has shown that 80 per cent of this volcanic lava outflow occurred during a period of magnetic reversal in the Earth's field, lasting approximately 10 000 years. Using the technique of argon-argon dating, the age of the lava was determined to be between 64 and 68 million years old, a time span straddling the K-T extinction.

Vulcanologists believe the source of iridium found by Alvarez in the clay K-T boundary layer to be the result of intense volcanic activity, which threw out huge amounts of volcanic core material into the atmosphere. Collecting airborne debris in filters placed 50 kilometres from the eruption of the hotspot volcano Mount Kilauea in Hawaii during the 1980s, scientists found particles enriched with iridium to levels 10 000 to 20 000 times higher than the normal Hawaii basalt rock. Until then, little was known about the movement of iridium from the Earth's core to the surface, but this data opened the door to an alternative explanation for the K-T event. Analysis showed rock derived from the Earth's mantle to have a similar mineral composition to that of stony meteorites and, on this basis, it was argued there was no need to postulate an extraterrestrial origin for iridium. Likewise, the formation of shocked quartz and mineral spheres at these sites was explained as a phenomenon occurring in the high-temperature, high-pressure conditions of an erupting volcanic core.

It is difficult to imagine the sheer size of volcanic upheaval that was required to lay down the Deccan Traps. In 1783, the eruption of Laki in Iceland released only 12 cubic kilometres of lava, but killed a quarter of the island's population and three-quarters of its livestock. But this was a puny event compared to the Deccan eruptions laying down an estimated two million cubic kilometres of lava, together with 30 trillion tonnes of carbon dioxide, six trillion tonnes of sulphur and 60 billion tonnes of the halogen gases chlorine and fluorine. The Deccan Traps was not the only region to be visited by volcanic activity and there is geological evidence of widespread explosive eruptions in many parts of the globe during this time period.

Such volcanic activity, over a long period of time, would be expected to cause a range of effects very similar to those proposed for an extraterrestrial impact. Huge clouds of volcanic dust encircling the Earth would reduce light and heat, leading to the equivalent of an impact winter; and the ejection of hot volcanic debris and gases would ignite forests and produce acid rain. In accounting for the K-T event, there is currently no absolute evidence tipping the balance completely in favour of either an impact hypothesis or a volcanic origin. Attempts have been made to marry these two theories with a proposal involving an asteroid impact triggering volcanic action by powerful shock waves fracturing and destabilising the Earth's crust. Although a good idea in theory, there is unfortunately little evidence to support it.

Much has been written on the effects of the K-T event on animal life, but remarkably little has been said about the consequences to the plant world. Plants did not escape entirely from the devastation visited upon animal life, but the effects appear to have been much more localised rather than global. Interestingly, the major area of sudden plant extinction appears to have been the southwestern region of the USA, close to the site posited for the meteor impact in Mexico. Extinction of plants at higher latitudes was much less evident and there appears to have been no blanket extinction in the southern hemisphere at all, but rather a slow reduction in the diversity of species, related perhaps to the initial climatic cooling. But how did the plant kingdom survive the K-T extinction relatively unscathed?

Plants survive on rather limited requirements and, given a supply of carbon dioxide, water, nitrogen and minerals, all is likely to remain well. In the face of hostile environmental conditions, plants have the ability to shed leaves or whole branches and die back below ground level in the form of rhizomes or underground stems. These mechanisms, together with the production of hardy seeds, ensure that plants can wait out adverse conditions. A further possibility is to migrate to a more friendly neighbourhood, attainable through the agency of seeds carried on the wind or transported by animals and water.

At the time of the K-T extinction, uncontrolled fires wrought havoc and destruction on the forests, particularly in the northern hemisphere, leaving a large gap for any number of species to fill. In the initial conditions characterised by high rainfall and a cooler climate, this niche was filled by a variety of ferns which, in time, gave way to low-diversity rainforest as temperatures increased. Flowering plants, which were already diversifying prior to the K-T event, now underwent a major expansion and the grasses first made their appearance on the world stage.

The birth of the grasslands

Geological evidence shows the grasses to have established themselves first in the southern hemisphere, in South America and Africa, some 10 million years earlier than their counterparts in north America. Although the burning of forests created a potential space for grasses, the climate was too wet to allow serious colonisation. One more event was required to establish the wide distribution and success of the grasses. This event was the formation of mountain ranges. Until that time, rain-bearing winds had free access to continental interiors, producing a climate of high rainfall and relatively lush vegetation. But the tectonic upheavals that resulted in the formation of the Rockies, Andes, east African Ridge Mountains, western European Mountains and the southern Alps of New Zealand created large rain shadows, leading to a much drier, cooler climate unfavourable for tropical or subtropical vegetation and rainforest, which died out as a consequence. The death of these species left a very large niche for plants that were capable of thriving in conditions of low rainfall. The grasses were in an ideal position to exploit this opportunity and their biological characteristics, including rapid reproduction and windborne seed dispersal, ensured their swift colonisation of the land.

Some plants die back
below ground level
to wait out
adverse conditions.

The fossil record of 20 Ma shows that vast areas of grassland were to be found on both sides of the equator. Below the forest line of northern Europe the great steppes spread eastward from the European plain as far as Mongolia, while in North America the prairies extended across the heart of the continent, from Canada to Texas. Below the equator, the pampas of Argentina stretched from the river Plate to the Andes and in Africa, the grassveld was established from the southern Cape Province to Zimbabwe. A somewhat higher rainfall in some areas created savannah grassland, a landscape still dominated by grasses but with a smattering of trees and shrubs. All of these grasslands were characterised by a climate of recurring drought.

It is difficult to envisage, millions of years before our own appearance on this planet, what richness and diversity of life was supported by these grasslands. Even in our own time, the earlier pioneers travelling west over the American prairies described columns of bison '25 miles wide', which took five days to pass a given point, an estimated number anywhere between four and 12 million animals. Even with no fertiliser other than dung, these arid grasslands were able to sustain a biomass greater than anything modern farming methods have been able to produce.

Adaptations for survival

Herbaceous plants grow from a structure called the shoot meristem, a specialised collection of cells that produce tissues of new stems and leaves. In most plants the growing point is at the tip of the shoot, but grasses are unique in the fact that growing meristems occur at the nodes rather than at the tip. This seemingly unimportant observation in fact reveals one of the major reasons for the success of the grasses. Throughout the world, millions of herbivores from addax to zebra chew their way through tonnes of grass every day. If the growing points of the grasses were located at the tips of the stems, the plants would simply disappear at the first bite. But, since the growing points are located just below ground level, they are protected from the attentions of teeth. It is solely as a result of this method of growth that the grasses form an ever-renewable source of food for grazers.

The exception to this is the ubiquitous goat which will eat every part of the plant. A walk through many an African village infested by goats will reveal a barren stretch of earth compacted to hard, dry mud, an effect heightened by the loss of trees felled for firewood. With little to hold the soil together, the arrival of the rains turns this concrete soil into a river of mud, leaving behind erosion gullies with great cracks and channels as the earth is washed away. The disappearance of the soil ensures there is no possibility of regrowth, making the goat and the axe a powerful combination in the conversion of once-fertile land to desert.

Fire, either deliberately lit as in modern times, or occurring naturally by lightning strikes, is a benefit rather than a hindrance to grasslands. It has been estimated that huge fires swept across the North American prairies every 10 years or so. Fire passes rapidly over the plains, leaving the protected grass shoots unscathed and taking only the old growth. Any woody plants that have established themselves are also incinerated, thereby removing a source of competition. It is common practice in Africa to fire grasses just before the rains, allowing fresh growth to occur without the encumbrance of the old, dead grass which is simply a passive competitor for space.

Although the Poaceae were quick to colonise the continental interiors, the aridity of the climate created problems. All green plants derive much of their energy requirement from the process of photosynthesis, which basically involves the chemical conversion of carbon dioxide to sugars in the presence of light and chlorophyll. Carbon dioxide is taken into the plant through small pores, the stomata, on the under-surface of leaves. These pores, however, have a second function, which is to allow water loss. The evaporation of water from leaves not only cools plants, just as sweating cools our skin, but essentially produces a vacuum pump by which water is drawn up the plant from the roots in a continuous column through specialised channels, the xylem, to all areas of the plant; even the leaves of the giant Californian redwoods grow at heights of 70 to 100 metres. This process, known as transpiration, works very efficiently for plants in moist, temperate climates, but the problem facing the dry, arid, tropical grasslands is that, if the stomata were to remain open all day to furnish the need for carbon dioxide intake, this would result in excess water loss, leading to desiccation of the plant. Various adaptations have been adopted by plants in desert environments but the grasses have solved the problem by evolving a new method of photosynthesis.

In the process of photosynthesis, carbon dioxide becomes fixed in a complex series of chemical reactions to form sugars. In most plants, the first intermediate molecule formed in this series of reactions has three

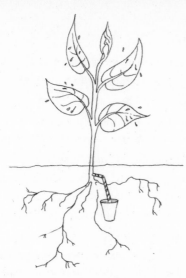

The evaporation of water produces a vacuum pump by which water is drawn up the plant from the roots in a continuous column to all parts of the plant.

carbon atoms (phosphoglyceric acid) and so this form of photosynthesis has been termed the C3 pathway. In about three per cent of plants, however, notably those growing in more arid conditions, for example tropical grasses, bougainvillea, asters and amaranths, a novel approach to photosynthesis has emerged. Here, carbon dioxide is taken up by the plant to form a four-carbon molecule (oxalo-acetic acid). Hence this is known as the C4 pathway. At a later stage, including the hours of darkness, the four-carbon molecule breaks down, releasing carbon dioxide into the tissues, which is subsequently utilised in the C3 pathway. In tropical grasses, C3 and C4 photosynthesis proceed side by side. The advantage to this system is twofold. Firstly the plant has an additional mechanism enabling it to take up substantially more carbon dioxide than it would using the C3 pathway alone. Secondly, and possibly more importantly, this extra capacity for carbon dioxide fixation allows the plant to function with the stomata only partially open, which leads to significantly less water loss than would otherwise occur. The advent of C4 photosynthesis was therefore an evolutionary step that prevented desiccation of plants living in arid conditions and, by virtue of better carbon dioxide binding, promoted rapid growth. Sugarcane and maize, both C4 grasses, fix carbon at twice the rate of wheat, a more primitive C3 grass. Using sophisticated techniques involving carbon isotope breakdown and evaluation of fossil tooth enamel (which is an indication of C3/C4 diet composition), it has been proposed that C4 plants evolved approximately 16 Ma, with a major expansion some seven to five Ma.

The grazing succession

Serengeti is a word derived from the Masai language, *Serengit,* meaning a sea of grass. The Serengeti Plains occupy an area of some 9 000 square miles in northwest Tanzania and extend across the Mara River into Kenya, where they are known as the Masai Mara. At any one time the plains may hold close to 750 000 animals, mostly grazers: wildebeest, zebra, Thomson's gazelle, topi and buffalo. The southeastern corner of the Serengeti is largely a flat plain of short grass and little rainfall. The animals concentrate here during the local wet season and migrate westward as the dry season sets in, to areas of higher rainfall and longer grasses. As this region dries out, a further migration to the north takes place to regions characterised by the highest rainfall and longest grasses. But long grass is a problem for smaller antelopes to deal with. Not only is it a physical obstruction preventing access to the shorter and more nutritious grass in the lower herb layer, but it is also poor in nutrients, being composed largely of cellulose and lignin with little protein. The community of grazers has adapted to this in a remarkable way.

The Serengeti is not a uniformly flat plain; it contains numerous undulations of varying height, drainage channels and islands of rocky outcrops known as kopjes or inselbergs. The distribution of long and short grasses on these undulations is determined by drainage, the shortest grasses being at the summit with the longer grasses found at the base where runoff water collects and persists for longer periods, allowing further growth. The wet season finds mixed groups of herbivores grazing the short grasses, which have a high protein content. At this stage there is adequate food for all, and often a surplus. As the dry season advances and the short grasses are consumed, the animals are forced to descend to areas of longer grass. The zebra possesses a digestive system able to cope well with longer grass containing less protein (see Chapter 4) and is the first to move down the slope to longer pastures. As it not only eats but also tramples down the long grass, the zebra prepares the way for the wildebeest to descend and take advantage of the medium-length grass more to its liking. And, in turn, the wildebeest prepares the stage for the smaller Thomson's gazelle by exposing the shortest grasses. This differential use of the various herb layers constitutes what is termed the 'grazing succession'. The clear advantage of this system is that it provides a way of supporting a number of different species in the same area without competition for a single food resource and is therefore able to sustain a much greater animal population than might otherwise have been possible.

The movement of game across the Serengeti is determined by rainfall, grass length and selectivity of the diet. Richard Bell and his co-workers have elegantly demonstrated this in the western Serengeti by accurately counting, on a daily basis, the number of different species on a series of transects 3 000 yards long and half a mile wide. Following the rains, the zebra are the first to move into an area, followed successively by wildebeest and Thomson's gazelle. Analysis of the stomach contents shows that, as expected, grass predominates in the diet of zebra and wildebeest but with a higher ratio of stem to leaf in the zebra (a feature typical of longer grasses). The diet of the Thomson's gazelle was found to contain almost 40 per cent fruit from dicotyledonous plants found in the lowest herb layer.

The human footprint

Sadly, the past 150 years have witnessed a progressive and relentless destruction of the natural grasslands by our own species. Faced with an exponential increase in human population the grasslands have been subjugated to the needs of cattle or have fallen under the plough. As long as humans exist, this loss will be permanent. Over millions of years the grasslands were able to support large numbers of animals, insects, birds and reptiles, with only dung as a fertiliser. The American buffalo or bison that once roamed the plains in numbers measured in millions can now be counted in hundreds, and only then in national parks. Likewise, the last refuge of African wildlife resides within the borders of its parks, a precarious state given the nature of African politics and the ever-increasing demand by humans for land. There is nothing today to match the original diversity of the natural grasslands and, without concerted effort, we are in danger of losing even this. So why is the world so anxious to poison pastureland with thousands of tonnes of fertilisers and pesticides, which are lethal to most forms of life and which are largely washed away into water courses where they can inflict further havoc on aquatic life? Responsible stewardship of the remaining grasslands requires our species to wake up fast and to consider the consequences of its actions. The axe, the goat, misguided agricultural policy and rank stupidity may yet make nomads of us all again.

CHAPTER TWO

HARVESTING
THE RESOURCES
the mechanics of feeding

Although the whole dental apparatus of an individual may spend the night hours in a glass, oral dignity can rapidly be restored the following morning with the aid of a little adhesive.

The origins and structure of teeth

So important are teeth in human society that whole sections of scientific endeavour – dentistry and orthodontics – are devoted to their care, maintenance and replacement, often at great expense to the consumer. Although the whole dental apparatus of an individual may spend the night hours in a glass by the bedside, oral dignity can rapidly be restored the following morning with the aid of a little adhesive. But without the luxury of dental science, a toothless person faces a diet restricted to very soft foods and liquids, with the possibility of progressive malnutrition in the midst of plenty.

Teeth and beaks form the first point of contact in the chain of events that converts food to usable energy. These sets of apparatus not only harvest the food itself but also prepare it for the process of digestion. Modification of the basic quadruped or avian blueprint is determined by the nature of the diet but, as we will see from the examples of the eagle and the duck, the commitment to a specialised food source also demands changes in a whole variety of other anatomical structures.

Before teeth evolved in any species, insects were able to cut through vegetation using their mandibles. Some insects, such as butterflies, could draw up nectar from flowers via a straw-like proboscis. The majority of animals were, however, required to be filter feeders, taking into a jawless mouth large quantities of water containing plankton or organic detritus, or else sucking in soil containing soluble nutrients. The development of teeth and flexible jaws represented a major evolutionary advance, making variations in diet and the construction of larger bodies possible. The one notable exception to this rule was the baleen whale, which, though remaining a filter feeder with no teeth, became the largest mammal of all!

The first appearance of a structure vaguely resembling a tooth came in a poorly understood group of early vertebrates known as conodonts. Living in the seas and freshwaters of the Devonian period, some 400 Ma, these small, soft-bodied, eel-like creatures possessed a jawless mouth containing two sets of mineralised dental tissue: anterior barbed ridges which could be pushed out of the mouth by a 'tongue' or cartilage plates to snare or impale prey. Once these ridges were withdrawn into the mouth, the trapped food was sliced by other, posterior elements. Freed from the necessity of filter feeding, the conodonts were active predators. Microscopic examination of the fossilised teeth of these animals has revealed a mixture of dentine, enamel and calcium phosphate, all materials found in present-day teeth.

The arrival of true teeth had to await the development of hinged jaws, a critical evolutionary event first seen in a group of fish called placoderms (literally meaning 'armoured skin'), which evolved in the Silurian period some 430 million years ago but which came to prominence only in the Devonian period. These unusual fish were mostly encased, apart from the tail, in heavy, bony armour with a prominent head shield of fused bony plates. Though little is known of their lifestyle, their heavy armour probably consigned them to a life at the bottom of the ocean where they were actively predacious. It was believed that teeth found in some members of this group were derived from the bony plates covering the head, or from external skin denticles at the margin of the jaws, but recent work by researchers Moya Meredith Smith and Zerina Johanson has established that placoderm teeth were formed with a central pulp cavity covered by dentine as in the true teeth of more modern-day vertebrates. Their research suggests that teeth originated independently within the placoderm group and also within other jawed vertebrates.

Teeth are a unique vertebrate acquisition and, regardless of function, each mammalian tooth follows a common architectural plan. The crown of the tooth projects above the line of the gum and is coated with enamel, the hardest substance in the body, composed largely of calcium phosphate and calcium fluoride. Beneath the enamel lies a layer of dentine, a substance similar in structure to bone, but much harder. Unlike enamel, which is not replaced as it wears, dentine is deposited on a daily basis throughout the life of the tooth. The innermost layer is composed of a pulp cavity containing connective tissue together with nerves and blood vessels which enter through the roots. Roots are inserted into sockets in the jaw and are covered in another bone-like material, the cementum. The whole tooth is anchored by bundles of tough collagen fibres which connect the cementum to the jawbone.

The mouths of mammals contain four basic types of teeth, which are modified according to diet. Incisors at the front are chisel-like in appearance and are used for cutting or clipping. Next to them are the canines, sharp and pointed for puncturing or grasping, and behind them are the premolars and molars for crushing and grinding.

Mammals: taking the gap

Living in the shadow of the dinosaurs, the early mammals were diminutive shrew- or weasel-like creatures, largely nocturnal in their habits, foraging for insects, grubs and worms in the leaf litter of the forest floor or seeking out fruits in the canopy high above. The K-T extinction (see Chapter 1) was to change this balance of power completely, leaving only the crocodile to represent the future prospects of a now-vanquished race of megareptiles. The sudden removal of competition allowed mammals to diversify and extend their range in ways previously closed to them; and, in the absence of large predators, building larger bodies was no longer as hazardous as it once might have been. But a larger body brings with it the need for specialisation and for a higher food intake. Limbs evolved along a variety of lines so that some species became diggers while others became runners, swimmers or climbers. And, along with these features, came different dietary requirements making it necessary to modify teeth and digestive systems for new and innovative methods of feeding. Teeth that are adapted for nibbling vegetation are not likely to be suitable for tearing flesh.

The emergence of the grasslands opened up a whole new range of possibilities for the life of mammals, providing a new home for herbivores with an ever-renewable food supply and a place where predators were more easily visible than they had been in the dense forest undergrowth. Plants could do little to outrun herbivores, but they could discourage them by secreting a variety of bitter-tasting or frankly poisonous chemicals. The grasses evolved to absorb silica from the soil, an abrasive mineral that induces rapid wearing of teeth. Therefore, in order to take full advantage of grass as a new food source, teeth needed to be structured in such a way as to minimise wear. The evolutionary response of grazing mammals was to develop molar teeth with a pattern known as hypsodont or 'high crowned'. When such teeth erupt they are initially covered with a layer of cementum, but as this wears away it exposes a complicated ridge pattern consisting of alternate layers of enamel and dentine. These two substances differ in hardness, wearing at different rates and thus assuring the production of rough surfaces at all times.

The protein-rich cytoplasm of plant cells is well protected by a cell wall composed largely of cellulose, a tough complex carbohydrate that is insoluble and largely immune to breakdown by digestive fluids. Faced with this problem, the initial approach of herbivores is to break down plant material mechanically as far as possible using the grinding and crushing action of molars and premolars. Anyone who has watched a cow or a goat eating will know that herbivore mastication is a side-to-side chewing action and this is reflected in the alignment of ridges in this plane on the molar teeth. Prolonged chewing is the rule in plant eaters and is essential for the disruption of plant cells. This dedication is further evolved in the ruminants which spend many hours regurgitating the stomach contents for further chewing. In most mammals the molar

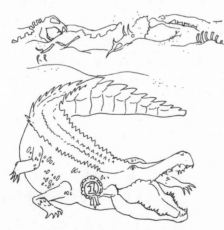

The K–T extinction was to change this balance of power completely, leaving only the crocodile to represent the future prospects of a now-vanquished race of megareptiles.

teeth are not replaced once the permanent teeth have erupted, but the elephant possesses six sets of molars which erupt in sequence throughout life, the final set appearing at about the age of 45 years. Once this last set has been worn down, the elephant confines itself to feeding on the softer grasses usually found in marshlands, but eventually it will succumb to malnutrition or predation.

Herbivore adaptations

Although individual species of herbivores may be recognised by the characteristic ridge patterns on the molar teeth, all molars are functionally identical. There does seem to be more room for versatility in incisors however, particularly in the elephant, beaver and hippopotamus.

Elephant tusks are the modified incisors of the upper jaw and function as surprisingly delicate tools acting as levers or probes. The bark of certain trees contains substances – fats and fatty acids – that appear to be essential to the well-being of elephants. In order to harvest this resource, elephants use a tusk to create a hole in the bark and then to undermine a whole section of it before using the trunk to strip off long segments. Unfortunately, this activity frequently results in trees being ringbarked, leading inevitably to the death of the tree. Areas with a high density of elephants can, in a fairly short space of time, sustain severe damage, resulting eventually in the degradation of woodland into savannah (see Chapter 12).

In its search for essential fats, the elephant seems to find the baobab or 'upside-down tree' to be a particular favourite. These remarkable trees may achieve an age of over 3 000 years and have an incredible capacity for storing water – differences in girth of over a metre have been observed between wet and dry seasons. Elephants seem to have an affinity to particular baobabs, worrying away at a tree for months or years while leaving others nearly completely untouched. The pulp of these trees, when fermented in the elephant gut, is thought to release alcohol, and animals often behave in a peculiar or aggressive manner after feeding on it.

Another essential requirement to the elephant diet is salt. The wise elephant matriarch guides her family group on a regular basis to a variety of salt licks where tusks are employed as picks or shovels to loosen impacted, mineral-rich earth before consuming it. Large pits and even caves may be excavated by such activity. This need for salt

In its search for essential fats, the elephant seems to find the baobab or 'upside-down tree' to be a particular favourite.

consumption is sometimes reflected in the risk elephants are prepared to take in order to find and consume it. In one instance elephants have been recorded following steep, narrow mountain paths before entering labyrinthine caves in pitch blackness to secure their prize.

Tusks also make formidable offensive weapons, notably in dominance contests between mature males, particularly if this involves a dispute over access to an oestrus female. In somewhat gentler fashion they may be employed to discipline rowdy teenagers causing irritation among their elders. Elephants have also learned that tusks do not conduct electricity and can be used to tear down electric fences, making it difficult to deny them access to almost anywhere.

In the hippopotamus, tusks are formed from the upper and lower canines plus the lower incisors. They play no part in feeding and are used entirely as weapons. Along stretches of rivers where hippo are plentiful, a mature bull may control a territory of no more than 50 metres or so, but will aggressively defend it against other males in fights, which often occur with little in the way of preamble and which frequently end with the death of one of the participants. This overt aggression reaches its peak during breeding season and it is mainly during this time that humans fall victim to hippos which can, and often do, bite small boats in half. Hippos are responsible for more human deaths in Africa than any other animal.

In contrast to the molar teeth of most mammals, which do not grow once they have erupted, the incisors of the elephant and the beaver continue to grow throughout life. The crowns on the beaver's incisors are constantly worn down by its tree felling and lodge building, but growth from the root ensures they are never worn away.

Carnivore adaptations

The teeth of carnivores are dedicated to holding, subduing, killing and tearing. Incisors are small and chisel shaped, used for stripping flesh from bones, while canines are well developed and sharp, designed to stab, hold or bite. Large felines typically use the canines to secure a throat hold, which causes death by asphyxiation as a result of compression of the trachea. An alternative is to take the whole muzzle of the prey animal into the mouth, likewise producing death by asphyxiation. When confronted with larger prey, cats may deliver a single bite to the spinal column, which severs the spinal cord, causing paralysis or sudden death. The dog family (wolf, African wild dog and coyote) have shorter canines than cats and cannot effectively employ a throat hold to subdue or kill prey. Instead, they tear open the abdomen, disembowelling the prey, or bite into selected regions such as the groin to effect rapid blood loss.

Once the prey is dead, these carnivores use their canines to tear into the flesh. Large chunks of meat often need to be 'cut' into manageable sizes before they can be swallowed, and this is the function of the carnassial teeth. Carnassials are elongated shearing blades formed by the combined action of the last upper premolar and the first lower molar. When the jaws are closed, these teeth slide against each other like the closing blades of a pair of scissors. Chewing takes place in a vertical direction, unlike the horizontal grinding action found in herbivores.

Dogs possess more teeth than cats, the consequence of an increased number of premolars. This design, coupled with more powerful jaws and reinforcement of the skull to anchor jaw muscles, permits them to break and fragment bones, an activity in which cats seem to have little interest (see Chapter 5). In general terms, cats are flesh eaters, concentrating on muscle and softer parts of a carcass, while dogs are non-specialists, eating almost everything apart from hooves and skull.

One of the enigmas of carnivore evolution is the emergence and subsequent extinction of the sabre-tooth cats. In these animals, the development of the upper canines and the carnassials was carried to extremes, but unfortunately there is little direct evidence from the fossil record to indicate with any certainty how they were deployed during feeding. Although it is tempting to say the huge canines were used to take large bites, examination of fossil mandibles indicates the jaw muscles were probably not strong enough to accomplish this. Perhaps they were used as slashing or stabbing weapons to vulnerable areas of their prey, causing rapid loss of blood. Being so large, the canines were probably used with great care to avoid damage to them, and a slashing bite to the abdomen may have been the preferred approach to killing, followed by a deep bite to the neck of an immobilised animal, which would quickly have wrecked major blood vessels. This much can be agreed upon, but there is no consensus on the reasons for the extinction of the sabre-tooths. Perhaps it was an over-commitment to a specific prey species which, for reasons unknown, became extinct, taking the sabre-tooths with it.

Birds – a profusion of bills

Birds have a high metabolic rate and the gift of flight comes at the cost of high energy expenditure. It is therefore a matter of necessity for birds to seek out food of high calorie value, preferably of high protein content, such as insects, arthropods, small vertebrates, fruit and seeds. In these circumstances, the digestive tract can be short, compact and of relatively light weight, an obvious advantage for flying.

In mammals, strict adherence to a purely vegetable or meat diet has, over millennia, produced dental arrangements well adapted to these types of feeding, with relatively little variation. With birds, however, nature seems more inclined to reveal her versatility. Consider the ways in which birds are able to go about the business of feeding:

- digging into the soil, sand or skin for worms, crustaceans or parasites
- breaking open hard seeds
- drilling into wood for larvae and insects
- catching insects on the wing
- cracking open large bivalve shells
- spearing fish or amphibians
- filtering invertebrates from water
- tearing apart meat
- sucking nectar from flowers
- using tools such as twigs to dig larvae from wood
- catching fish in a specialised pouch

Needless to say, not all these options are available to all birds. If they were, a bird's beak would require more attachments than a Swiss army knife. There is, however, some room for versatility – a beak capable of harvesting insects and worms in summer may be suitable for cracking open seeds in winter.

The possession of a highly specialised bill does not occur in isolation, but forms part of a package that involves modification to a number of structures in the basic avian model, the details of which commit the bird irrevocably to a certain way of life. Consider, for example, an African eagle drifting slowly over the grasslands or savannah. It is searching for rodents, lizards, birds and the young of small antelopes or monkeys. To do this requires spending much time aloft while expending little energy on active flight. In other words, it has to glide. Wings designed for gliding enable it to make judicious use

It is a matter of necessity for birds to seek out food of high calorie value.

of thermal air currents: a large wing area provides good lift and this is augmented by slotting of feathers at the wing tip. Soaring high above the plains, the eagle needs not just good, but superb, high-resolution binocular vision. The importance of vision to all birds can be inferred from the fact that the eyes constitute 15 per cent of the weight of the head, compared to only one per cent in humans. In birds of prey, the eyes are forward facing, a requirement for highly refined binocular vision. Visual acuity – the ability to distinguish small objects close together at a distance – is considerably greater than in humans and is achieved by a much higher density of rods and cones in the retina. Once the eagle has selected potential prey on the ground or in the air, it then needs to render it immobile. The feet are thus adapted to form sharp talons which not only grasp but which may perforate skin. Only when all these specialised anatomical structures have played their part, does the typically sharp, curved beak of the raptor come into play, by tearing into flesh.

These adaptations would, of course, be useless to the average duck, which needs a suite of quite different characteristics in order to occupy successfully its own particular niche in bird society. The beak of the duck is broad and flat, making it well suited for straining invertebrates from mud. The eyes are placed more laterally on the sides of the head; this produces a very wide visual field and allows the duck to see behind it, an arrangement typically found in species that fall prey to other creatures. Adaptations to an aquatic lifestyle include webbing of the feet for more efficient locomotion in water and an oily secretion over the feathers to render them waterproof. This oil is produced by the uropygeal gland at the base of the tail and is spread over the feathers during preening. Waterlogged feathers would otherwise prevent flight and lead to excessive loss of body heat.

The seed eaters are confronted with the problem not only of removing seeds' tough outer husk, but also of dealing with the hard kernel. Some birds, such as doves, swallow seeds whole, which then pass into a crop – a chamber modified from the lower end of the oesophagus – where they are softened by soaking in water (see Chapter 4). Further breakdown is then achieved in the gizzard, which is a modified part of the stomach, characterised by thick muscular walls and a horny lining. In some species, breakdown of seeds by the churning action of the gizzard is facilitated by their swallowing small stones. Finches, on the other hand, are able to remove seed cases by either crushing or cutting them. Cutting involves using the tongue to engage the seed in grooves present on the hard palate. By rapid forward and backward movements of the sharp edges of the jaws, the husk is effectively sawn through.

Hummingbirds and songbirds insert their beaks into flowers to draw nectar up from the nectar chamber using their tube-like tongue as a drinking straw. They are specialist feeders: the length and curvature of the bill corresponds precisely to the configuration of the preferred flowers. In order to probe flowers with a high degree of accuracy, they have evolved a unique method of hovering flight. During hovering, the body is held in an almost vertical position. On the first downstroke, the wings move forward and downward as would be the norm for routine flight, but the wings then rotate at the shoulder joint during recovery so that the next downstroke is directed downwards and backwards. This is repeated at the almost unbelievable speed of up to 80 beats per second, and keeps the bird in a fixed position.

The Makgadikgadi saltpans in north central Botswana are remnants of a large inland lake which ultimately dried out as a result of tectonic activity diverting the courses of the rivers that fed it. The resulting pans now contain shallow water only in the southern African summer, from December to April, at which time they provide a home for over one million flamingos. The alkaline water is essentially a warm broth of small invertebrates and blue-green algae, the perfect diet for flamingos who migrate up and down Africa's rift valley to plunder this larder and to breed. Uniquely, flamingos feed with head and bill in an upside-down position, straining food particles from water or mud by moving their fleshy tongue back and forth, and so creating a suction pump. The upper and lower mandibles contain sets of matching lamellae for trapping particles in the water. The type of lamellae determines the diet: the coarser filters (large hooks) of the greater flamingo filter out small invertebrates, such as shrimps, for consumption, while the finer filters of the lesser flamingo trap blue-green algae.

How this alchemy is achieved, whereby diet and anatomical structure have developed symbiotically, is the business of evolution, and there is no better example than to follow the steps by which a diminutive creature of the woodlands became the tall proud master of the grasslands – read about it in Chapter 3.

CHAPTER THREE

CHARLES DARWIN, HORSES AND FINCHES
understanding evolution

Darwin proposed that new species arose by modification of existing species.

A new view of the world

On 24 November 1859 a book appeared that would change forever the way we see the world. The first edition of 1,250 copies was sold out on the first day. The work in question was *The Origin of Species* by Charles Darwin MA, who had previously written of his experiences as a naturalist on board HMS Beagle during its five-year journey of exploration around the world.

Darwin's thesis contained two separate but interrelated ideas. The first of these was the assertion that species are not fixed, but rather in a continuous state of flux. Variations occurred between individuals and were related in part to inherited characteristics and partly to random chance, what we would now call mutation. He proposed that new species arose not from separate acts of creation but by modification of existing species. This view arose from a multiplicity of observations made during his travels, but probably the most influential were those recorded during his visit to the Galapogos Islands. Lying in the Pacific Ocean some 950 kilometres west of the Coast of Ecuador, these islands are well and truly isolated. The 25 islands that make up this cluster were formed separately from volcanic activity on the ocean floor. They have never been joined together and their isolation from each other is compounded by strong, treacherous currents. Any movement of fauna between islands is thus largely impossible. The Governor of the day was able to point out to Darwin the variations encountered in the shells of land turtles living on the islands, and so specific were they that the examination of an individual shell would be sufficient to identify the origins of its owner. This remark seems to have passed Darwin by at the time and it was only years later that its true significance became apparent.

As part of his collection of specimens, Darwin brought back to England a number of finches that he had gathered on various islands. It is known that finches strongly resist crossing open water. The ancestors of those on the Galapagos Islands are therefore assumed to have been blown there from South America, probably by hurricanes, and to have evolved independently on different islands. Once again, Darwin seems to have given his finch specimens little thought beyond their obvious similarity, until a meeting with John Gould, a British ornithologist, convinced him that the birds were in fact distinct species. Possibly at this time he recalled the story of the tortoise shells.

It seemed perplexing that each species, whether tortoise or bird, could have arisen *de novo* by wholly separate acts of creation and lines of development, and yet be so similar. But if he accepted the possibility that each species resulted from changes to the ancestral stock carried over on the currents and winds from South America, then similarity and diversity would be expected. In this idea was the germ of a theory of species modification across successive generations. Against the philosophy of his time, Darwin accepted the reality of evolutionary change but was lost for a mechanism that would bring about this change.

Two years after the Beagle had safely docked back in England, Darwin came upon an article entitled *Essay on the Principle of Population* by the Reverend Thomas Malthus. This was a rather depressing document dealing with human population growth, in which Malthus concluded that, if left unchecked, the population would expand in geometric fashion to a point where it would outstrip its available food supply. It was only the intercession of war, disease, famine and natural disasters that prevented this. Herein lay the solution to Darwin's problem, and he was quick to grasp the possibility that the same principles of population may be applied to all living organisms. In the face of limited resources, and hence competition, those with structural features better adapted to the prevailing conditions would survive and flourish and, in time, would outnumber or even displace those who were not so adapted. In Darwin's own words:

'if variation useful to any organic being do occur, assuredly individuals thus characterised will have the best chance of being preserved in the struggle for life; and from the strong principle of inheritance they will tend to produce offspring similarly characterised. This principle of preservation I have called, for the sake of brevity, Natural Selection'.

The consequence of natural selection is that, over many generations, the character and composition of any given species will change. Some morphological features will become more prevalent and may, in time, be modified to such an extent as to bear little apparent resemblance to the ancestral stock, for example the evolution of birds from reptiles.

Natural selection will be influenced not only by availability of food resources but also by other agencies including climate, geographical barriers or isolation and type of environment. Darwin was well aware of the ramifications of his proposals in an age when a powerful church insisted on a literal interpretation of religious dogma and where no less a person than Dr John Lightfoot, the Vice Chancellor of Cambridge University, had calculated from biblical references that the human race was created at nine o'clock on the morning of 23 October 4004 BC. Not only would the concept of evolution demolish the literal interpretation of the *Book of Genesis*, but it would also topple mankind from its self-proclaimed status as a direct creation in the image of God. Little wonder that for many years Darwin kept his ideas locked away in a drawer, along with a letter to his wife giving instructions to publish in the event of his untimely death.

Eventually persuaded to publish, Darwin had completed preparation on the first 10 chapters when he received a manuscript from the East Indies entitled *On the Tendency of Varieties to Depart Indefinitely From the Original Type*. The author was Alfred Wallace and, much to Darwin's surprise and probable mortification, it was a complete précis of his own ideas – 'the struggle for existence' and 'the constantly changing character of species'. Darwin was later to describe Wallace as 'generous and noble' while Wallace dedicated his own book to Darwin as a token of 'personal esteem and friendship'. The joint discoverers of evolution thus became an object lesson in graceful manners, a form of behaviour almost extinct among modern academics.

Phyletic gradualism

The theory of evolution originally proposed by Darwin and Wallace, of structural change by descent with modification accomplished by natural selection is known today as 'phyletic gradualism'. Inherent in this view is the assumption that the fossil record should reveal a succession of intermediary forms as one species evolves into another. A case in point is the ancestral Archaeopteryx, a pigeon-sized creature showing a mosaic of reptilian and bird features, originally unearthed from Jurassic deposits in Germany only two years after the publication of *Origin of Species*.

Living in the ancient forests of what is now North America some 60 million years ago was a small mammal no larger than 35 centimetres in height, browsing on the succulent leaves of a variety of shrubs. To suit this diet it possessed short crowned teeth, somewhat resembling those of pigs. Its short legs had four toes on the forefeet and three on the hind feet. This was the Dawn Horse (*Eohippus* or *Hyracotherium*), living at a time when the first grasses were starting to appear but where the landscape was largely dominated by forest.

Thirty million years later, the interiors of the large continental masses became significantly more arid, an environment much better suited to grasses than trees. At this time, the descendant of *Eohippus*, now called *Mesohippus*, had appeared. It stood 45 centimetres tall and its longer legs had three toes on each foot. Its teeth were modified to include premolars or incisors, which were capable of chopping and grinding, giving it access to a wider variety of vegetation, but it was still basically a browser. A further 10 million years were to pass before the next 'generation' *Merychippus* stepped out on to what must have appeared to be infinite grassland. The weight of each foot was now essentially placed on a single toe, the remaining two having become vestigial. The teeth were high crowned and had developed dentine and enamel to reduce the wear and tear of grinding. The neck had also lengthened, allowing for better surveillance across the grasslands and also the ability to eat at ground level without the need to kneel on its front legs.

The next stage was *Pliohippus*, which was the prototype of the modern horse as well as being the ancestor of the zebra and ass. This appeared approximately six to seven million years ago and was equipped with feet of a single hoof supported by strong ligaments that gave it increased mobility and speed. Both incisors and molars were firmly adapted to a diet exclusively of grasses.

The evolution of the horse is cited as the classical example of phyletic gradualism, but it is important to point out that this evolutionary pathway did not occur as the result of a simple linear progression – like climbing a ladder. Natural selection modified a number of the equine ancestral stock in several ways with variable success, some becoming extinct in a relatively short period of two million years while others survived for almost 20 million years.

Central to Darwin's thesis was the idea that natural selection acted on individuals through the medium of an inherited package derived from both parents. At that time, the nature of such a package was unknown and was visualised along the lines of the free mixing of paints of different colours. It would be years before the Austrian monk, Gregor Mendel, defined the rules of inheritance in his classical experiments with peas and beans, and even further in the future before the role of chromosomes and DNA were appreciated. According to natural selection, changes passing from one generation to the next are likely to be small and hence of limited immediate significance. However, a sudden change in the local environment may elevate these apparently insignificant features to a level that proves critical to the survival of the species. Appropriately enough, an example of this was observed on the Galapagos Islands in the 1970s.

Daphne Major is one of the islands in the Galapagos group. Scientists working on long-term projects on this island have been able to capture, measure, mark, release and recapture every single Medium Ground-Finch on the island, a considerable undertaking. In 1976, the average beak

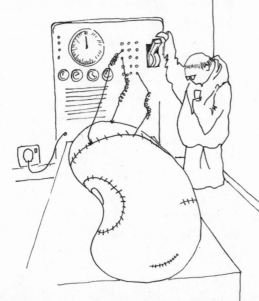

The Austrian monk, Gregor Mendel, defined the rules of inheritance in his classical experiments with peas and beans.

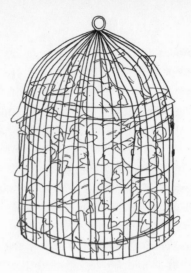

Remarkably, scientists have captured, marked and released every single Medium Ground-Finch on Daphne Major island.

length of the 741 ground-finches measured 9.4 millimetres, with a range varying between six and 12 millimetres. The following year witnessed a drought, when only 20 per cent of the expected rainfall occurred. Under these conditions, the number of new seeds produced by plants fell substantially and those that were produced were 50 per cent larger and harder than normal. As a consequence, all but one of the 388 hatchlings for that particular year died, together with most of the adult birds. Survival in the adults was directly related to beak length, because only those with larger beaks could crack open the bigger, harder seeds. In total, 91 per cent of birds with the average beak length (9.4 millimetres) died, compared to 60 per cent of birds with larger beaks. Survival came down to possessing a beak length only 0.8 millimetres larger than the average. If drought conditions had persisted, there would have been a permanent shift in beak length in favour of longer beaks – demonstrating how sudden and unpredictable environmental change can exert intense selection pressure on a species.

If natural selection acts on individuals, Darwin was left with the difficult task of explaining the origin of different species – a topic which, despite the title of his book, he considered only briefly. In his view, new species arose after countless generations had been subjected to minute incremental changes. Our attempts to divide living things into species somewhere along this continuum became an essentially arbitrary exercise, only deemed necessary when descendants were so different in appearance from the ancestral stock as to demand a new name. But Darwin himself was well aware that linear progressions in this fashion could not account for the great diversity presented by nature.

Allopatric speciation

This problem about the origin of species was not seriously addressed for another 90 years or so until the geneticist Theodosius Dobzhansky published his work Genetics and the Origin of Species in 1937. This work was subsequently extended by Ernst Mayr, a zoologist, in 1942. Mayr proposed that genetic variation would be more likely to occur if populations of a particular species were to become geographically isolated from one another, with one group being exposed to different ecological settings and different environmental pressures acting on natural selection. Geographical isolation could be brought about by a number of mechanisms, including changes to a river course, the intrusion of desert, advancing glaciation or mountain ranges, the appearance of new land bridges between continental masses or by organisms being carried across oceans on mats of vegetation. The proposed evolution of new species brought about by geographical isolation became known as 'allopatric speciation' (Gr. *Allos* = other, *patra* = native land). But how does it work?

Everyone is familiar with the slight structural variations within any given population of an animal or plant – fingers and beaks a little longer or shorter, stems or bodies somewhat taller or shorter but, undeniably, they represent the same species. If this were not so then evolution could never occur, because natural selection would have nothing to act upon. When a small population is cast adrift in a new environment, survival will depend in large measure on chance and whether natural selection has already (but unintentionally) equipped some individuals with the means to accommodate their lives more successfully to this new environment. Many individuals not so blessed will perish. Obviously, those better endowed will survive to reproduce, and natural selection acting on future generations will produce individuals even more appropriately adapted to the new living conditions, ultimately resulting in an organism significantly different from the ancestral stock – in other words, a new species.

Having found a mechanism that could adequately explain speciation, the next obvious question was how long might it take for one species to evolve into another? According to classical Darwinian theory, minute changes in structure are transmitted and amplified over countless generations and, on this basis, several intermediate forms should be apparent in the transition from ancestral parent to new species. Although several such progressions have been identified, the fossil record seems to be composed largely of periods of rapid change, with the emergence of clearly established new species, followed by long periods of stasis, where little, if any, morphological change occurs. Rapid change in this context is to be understood on a geological timescale, but nevertheless appears to be measured in millennia rather than millions of years. Darwin attributed these discrepancies to the incomplete nature of the record itself. Given the difficult conditions necessary for fossilisation and our own very limited geological exploration, it is hardly surprising that major gaps existed in the fossil record and, indeed, continue to exist.

Punctuated equilibrium

In 1972 two American scientists, Niles Eldridge and Steven Jay Gould proposed an alternative theory to Darwinian gradualism. They accepted the incomplete nature of the fossil record but were convinced it nevertheless represented the true state of affairs – that evolution was characterised by the relatively rapid appearance of new species followed by long periods of stasis culminating ultimately in inevitable extinction. The new theory became known as 'punctuated equilibrium'. It is important at the outset to emphasise that this new viewpoint did not detract from the essential core of evolutionary theory – modification by descent through natural selection – but was more concerned with the timescale and pattern of evolution. But what mechanisms could be responsible for such rapid evolutionary change?

We know that the appearance of a new species generally occurs when a small cohort of a particular organism becomes geographically and hence reproductively isolated from its fellows. The key word here is small, because genetic variations are much more likely to gain full expression in such a population. Within a large group, numerous random matings have the effect of buffering genetic variation because rapid dilution of the effect reduces the probability of wholesale change. The potential for rapid change in a small population can be brought about by genetic events such as gene deletion, gene substitution, mutation and the emergence of hybrids. But an environmental crisis may also be a powerful driving force in initiating rapid change by exerting intense selection pressure – as in the case of the Ground-Finches of Daphne Major. Included in such crises are periodic mass extinctions. Here, there is no better example than the immense diversity of mammals that appeared shortly after the K-T extinction, which removed their major source of competition, namely the dinosaurs – until then, the undisputed kings of the food chain.

Although now an established feature of most textbooks, punctuated equilibrium continues to attract criticism and controversy. The rapid emergence of a new species cannot be complete within one or two generations and, of necessity, must pass through intermediate forms. Where are the fossils to validate this rapid transit? Unfortunately, in this regard Eldridge and Gould find themselves in the same boat as Darwin, intermediate stages having demonstrated an exasperating refusal to subject themselves to fossilisation and/or discovery. Adding to this problem is the sheer difficulty of correctly interpreting the fossil record. A report published in 1995 reviewing a large number of studies that had specifically set out to investigate whether the fossil record predominantly favoured evolution by phyletic gradualism or punctuated equilibrium concluded:

> *'Evidence overwhelmingly supports the view that speciation is sometimes gradual and sometimes punctuated and that no one mode characterises this very complicated process in the history of life.'*

What's in a phrase?

History shows that sometimes a single phrase may have far-reaching consequences neither intended nor anticipated by the author. One such example taken from Darwin's *Origin of Species* relates to 'the struggle for existence', usually paraphrased as 'survival of the fittest' – possibly the most misunderstood concept in 19th and 20th century biological science. The ramifications of this misunderstanding have sometimes been farcical and sometimes criminal but, before they are explored in any detail, it is worth looking at what Darwin actually said on the matter:

> *'I use this term struggle for existence in a large and metaphoric sense including dependence of one being on another, and including (which is more important) not only the life of the individual but success in leaving progeny. Two canine animals, in a time of dearth, may be truly said to struggle with each other which shall get food and live. But a plant on the edge of a desert is said to struggle for life against the drought. As the mistletoe is disseminated by birds its existence depends on birds; and it may be metaphorically said to struggle with other fruit bearing plants, in order to tempt birds to devour and disseminate its seeds rather than those of other plants. In these several senses, which pass into each other, I use for convenience sake the general term of struggle for existence.'*

From the outset, Darwin was careful to emphasise that the struggle for existence should be read as metaphor rather than a physical biological process. An ecosystem functions in many respects as a single unit where one form of life is dependent on another for its very existence and, in this sense, 'co-operation' might be a more appropriate key word rather than 'struggle'. But, unfortunately, it is the image of two dogs engaged in battle for possession of a single item of food that became the focus of attention, and has been responsible for emphasising what many have deemed the unacceptable face of nature, where violent competition becomes the normal pathway in achieving any given goal. Hence, we have become familiar with expressions such as 'a dog-eat-dog world' and also Alfred Lord Tennyson's familiar epithet of 'nature red in tooth and claw'. Darwin's scientific contemporaries, such as Thomas Henry Huxley, were quick to point out that nature, having no moral or ethical sense, could never be accused of wanton cruelty. These higher values were essentially human and we, by virtue of culture and social organisation, could transcend the inherent nastiness 'out there' in the wild world – noble sentiments showing that presumably Huxley had momentarily forgotten the barbarity of the Spanish Inquisition and the Crusades as well as numerous Imperial massacres in the name of colonial acquisition.

War, the subjugation of one people by another and various forms of racial prejudice have been part of the human story almost from its beginnings. It is to be much regretted that, in relatively recent times, the concept of a 'struggle for existence' became equated with the idea that evolution advances by conquest. Needless to say, this interpretation was eagerly seized upon by a variety of powerbrokers and megalomaniacs the world over as justification for gross abuses of human rights during colonial conquest, exploitation of the poor by the rich, the persecution of those of differing religious beliefs, etc. It also encouraged the rise of eugenics, the belief in a natural superiority of one race over another.

Rationalisation for murder

On the 20th January 1942, 15 senior officials of the Nazi party gathered at a beautiful villa in the Wannsee suburb of Berlin. The meeting was chaired by Reinhard Heydrich; Adolf Eichmann was assigned to take the minutes. On what Eichmann later described at his trial as a 'largely successful day' the group made substantial progress in its efforts to work out the logistics and practicalities involved in 'the final solution to the

Jewish question' – a rather quaint phrase for the proposed murder of 11 million people. Almost one-third of the Wannsee Protocol concerned a detailed discussion defining which Jews were to be automatically 'evacuated to the east' (a euphemism for extermination) and which could remain unmolested in the Reich. The categorisation of those destined for evacuation is a testament to Hitler's bizarre grasp of genetic principles.

All full Jews were to be automatically evacuated. Half Jews (*Mischlinge* of the first degree) were also earmarked for extermination unless considered to be providing essential services to the government. But they were to undergo 'voluntary' sterilisation as a precondition to remaining in the Reich. Quarter Jews (*Mischlinge* of the second degree) were to be treated as Germans unless mixed parentage, physical appearance or unacceptable behaviour were sufficient to identify them as Jews. These criteria applied to single people, but the stakes were considerably higher for those who were married. In this case, all those with Jewish blood, regardless of degree, were to be evacuated along with any progeny. The only exceptions were to be persons married to full Aryans – and only here because of the likely fuss created by the German relatives.

Hitler developed much of this philosophy during the 1920s, arguing that Germany, once a proud and vigorous nation, had lost the 1914–1918 war because the pure Aryan bloodstock had been undermined, diluted and contaminated from external sources, reducing the country to a shadow of its former glory. It has to be said that Hitler was partially influenced in his thinking and misappropriation of Darwinian evolution by likeminded eugenicists in the United States of America and Britain. In this regard, it is salutary to ponder the decision of the United States Supreme Court in 1927, which permitted the compulsory sterilisation of persons judged (often by unqualified personnel) to be mentally retarded.

Hitler was partially influenced in his thinking by a misappropriation of Darwinian evolution.

Creationists versus evolutionists

But, just as the unfortunate phrase the 'struggle for existence' was subverted to the use of racists and madmen, it was also hijacked by many leading churchmen. In part a reaction to the histrionics of the eugenics brigade, there was a more fundamental concern: a refusal to accept that God had created a world where slaughter and violence were the norm. The front line of this movement was represented by the Christian fundamentalists of America's 'Bible belt'. Committed to a literal interpretation of the Bible where the creation of the world and all living species took just 6 days, some states, notably Tennessee, banned any teaching of Darwinian evolution. In fact, the law prohibiting the teaching of evolutionary theory remained on the Tennessee Statute until 1967. Amazingly enough, the state of Arkansas was challenged in the court as recently as 1981 for its policy of insisting that equal time be given in schools to biblical creation and evolution. It was not until 1987 that the Supreme Court finally allowed the unrestricted teaching of evolutionary theory in the United States of America.

In retrospect, Darwin probably reflected on the wisdom of his use of the 'struggle for existence' and the imagery evoked by the fighting dogs, despite his insistence on the metaphorical nature of both. It seems unfair to accuse him of naivety, even though the fires he helped to light under the church and the devils of eugenics continue to burn today.

The evidence for evolution is irrefutable but, regardless of religious dogma or the claims of science, creation myths have formed a central part of all human cultures since their beginnings. They are powerful, lyrical and evocative stories that fulfilled an archetypal need and presented us with a rationale for existence and an anchor in the infinite space of a seemingly indifferent universe.

The image of two dogs engaged in battle for food has emphasised what many have deemed the unacceptable face of nature.

CHAPTER FOUR

MAKING A MEAL OF IT
diet and digestion

In some species (the ruminants or foregut fermenters),
the stomach is partitioned into four distinct chambers.

The plant eaters: breaking into the storehouse

Insects and animals have been eating plants for over 250 million years. By any standards this represents an impressive quantity of vegetation. Which makes even more remarkable the fact that, in all this time, no multicellular organism has ever possessed a direct means to effect the chemical breakdown of cellulose, the main constituent of plant cell walls, and guardian to the protein-rich cytoplasm. On the surface this seems to be a curious omission, a failure to provide what surely must be an evolutionary imperative, without which the plant world could never be harvested as a food source. Was it really possible that the synthesis of an enzyme capable of doing the job represented a bridge too far? Nature chose to solve this particular problem by travelling an indirect route, conscripting the services of certain micro-organisms – bacteria and protozoa.

Cellulose is a complex polysaccharide, which means it is composed of a large number of sugar molecules joined together with chemical linkages into a long chain. In order to convert this to simple, high-energy sugars, it is necessary to fracture these links. The bacteria in the gut of all plant eaters have the capacity to do precisely that, through the agency of an enzyme, cellulase, which only they can produce. In exchange for their unique biochemical gift, the micro-organisms are provisioned by their host with the materials necessary for their own existence. Of necessity, this symbiotic (Gr. *symbiosis* means 'living together') relationship must have existed from the earliest times. It was critical to the diversification of land animals, who previously were restricted to a diet consisting in the main of organic detritus, but who now were able to exploit an infinitely renewable food source and to gain possession of high-energy molecules, enabling them to construct larger, more active bodies.

The chemical conversion of cellulose does, however, take time and, in order to increase efficiency, the mammalian herbivores have developed modifications to parts of the digestive system to form chambers where the breakdown of cellulose can be facilitated. In some species (the ruminants or foregut fermenters), this has involved partitioning the stomach into four distinct chambers, while in others (the caecal or hindgut fermenters), the large intestine is extended to form a blind sac, the caecum.

Ruminant digestion

The ruminants are represented in the wild by buffalo, wildebeest, giraffe and a wide variety of gazelle and antelope and, on the domestic front, by cattle, goats and sheep. As we have seen, mechanical breakdown of grass and foliage occurs by the grinding action of the molars and premolars. The food bolus is then swallowed, mixed with copious amounts of saliva, which not only contains water and mucus to ease its passage down the oesophagus, but also enzymes, notably amylase which breaks down starches into sugars. Food enters the first compartment of the four-chambered stomach, the rumen, which is a thin–walled, muscular sac lined with numerous corrugations designed to increase the surface area available for absorption. This chamber and its neighbour, the reticulum, function as a single unit, both containing the micro-organisms necessary for the breakdown of cellulose. The rumen acts as a fermentation vat where rhythmic muscular contractions mix the digested food with the bacterial 'broth'. In large herbivores, such as domestic cattle, which feed on hay and grass, the rumen may have a capacity of up to 200 litres. Within the fluid of the rumen, small particles of food sink while large, undigested plant fibres with low specific gravity float to the surface and are regurgitated back into the mouth for further mechanical breakdown by chewing, a process termed rumination ('chewing the cud'). Interestingly, brain wave patterns recorded during rumination are very much like those of natural sleep. Although time consuming – the process may occupy a period of several hours – rumination is nevertheless highly efficient and is designed to extract the maximum benefit from what may often be poor-quality food. The organic acids produced by the fermentation of cellulose are absorbed directly from the rumen and contribute up to 70 per cent of the animal's energy requirements.

The breakdown of cellulose, and hence of cell walls, releases the internal cytoplasmic contents of cells, which now become available for digestion. Proteins, like cellulose, are also subjected to bacterial fermentation, producing a variety of nitrogenous compounds including urea and ammonia, which are potentially toxic to the animal. As it happens these chemicals are the ideal building blocks from which the bacteria can construct their own proteins. Remarkably, this simple recycling of nitrogen has allowed herbivores to evolve a neat trick in dealing with the waste products of their own metabolism.

One of the principal waste products of metabolism in mammals is urea, normally excreted from the body as a dilute solution in the form of urine. Even with highly concentrated urine there is necessarily the loss of a certain minimum volume of water. By 'donating' the urea to bacteria resident in its own gut a herbivore can clearly conserve some of the water it would otherwise have lost as urine. In temperate climates where access to water is likely to be guaranteed, this may be of relatively little importance but for animals living in arid or desert conditions such conservation of water may be a critical factor in ensuring survival.

The bacteria resident in the gut of ruminants and non-ruminants alike are essential for life – indispensable for the breakdown of cellulose and the recycling of nitrogen. But in ruminants this association is carried one stage further – in what appears to be a nasty double-cross, the bacteria themselves are digested in large numbers further down the intestine to provide a secondary source of high quality protein.

The fermentation process generates a great deal of methane gas as a by-product, which is removed by belching. It has been estimated that ruminants worldwide contribute up to sixty tonnes of methane per year, the second major source of atmospheric methane after natural vegetable decay. Methane is one of a number of 'greenhouse' gases which contribute to the phenomenon of global warming. By preventing normal heat loss through the atmosphere, these gases are becoming a' major factor in causing substantial worldwide changes in climate. This matter was taken with such seriousness in New Zealand that the government proposed legislation that would have seen the introduction of a 'flatulence levy'. How this would have reduced methane emission remains unclear, but farmers, who were likely to bear the brunt of this new tax, were not slow in voicing their opposition and spawned a dissident group – F.A.R.T. (Fight Against Ridiculous Taxes). In an unusual victory for common sense, this project has now been abandoned.

A young ruminant, fed solely on its mother's milk, has no need of a functioning fermentation vat and, in these circumstances, the rumen is contracted down to form a passage which allows direct flow of milk from the oesophagus to the true stomach. On weaning, the young calf is required to prime its gut with the bacteria necessary for a vegetable diet – which it does, quite simply, by eating the faeces of an adult animal.

When fermentation in the adult beast has, after several hours, been completed, contraction of smooth muscle in the rumen and reticulum forces the partially digested food into the fourth chamber, the abomasum, which is the true stomach. The third chamber, the omasum, functions as a valve which strictly regulates the flow of material into this true stomach. Once in the abomasum, digestion proceeds along the general pathway established in mammals.

Caecal fermentation

In non-ruminants, fermentation of cellulose occurs almost at the end of the digestive tract rather than at the beginning. This group, the hindgut, or caecal, fermenters, is represented by the zebra, ass, rhino, hippopotamus, elephant and rodents in the wild and, domestically, by the horse and swine. Adjacent to the point where the small intestine joins the large intestine (colon) in humans is a small pouch called the caecum. It is intrinsically part of the colon and functions to absorb water and mineral salts from the remnants of the digestive process – digested food, gut secretions and the sloughed off cells from the upper reaches of the digestive tract. In hindgut fermenters, the caecum is modified into the form of an elongated blind-ending sac. The caecum, together with an enlarged colon, become the fermentation chambers required for cellulose digestion.

In these structures fermentation proceeds along lines similar to the ruminants – the same biochemistry, same micro-organisms, same breakdown products and same cycling of urea and ammonia. The overall efficiency of the process is, however, not as good as that seen in the ruminants. Firstly, the churning action of the rumen allowing thorough mixing of bacteria and food is absent in the caecum. This leads to incomplete breakdown of cellulose and produces a much coarser dung in which intact or partially intact plant material is evident. Some animals have dealt with this particular problem by eating their own faeces, a phenomenon known as coprophagy, which allows a second chance of digestion. This practice is common among rodents and rabbits and has also been observed in elephants, who may eat the dung of other elephants. It does not occur, however, in horses, zebras or asses. Secondly, the high-protein 'meal' provided by digestion of micro-organisms in the stomach and small intestine of ruminants is not available to the caecal fermenters and these organisms are simply lost in faeces.

You would be forgiven at this point for querying the reasoning behind placement of the fermentation system at the end of the digestive tract where its efficiency is always going to be less than optimal, but commitment to this system by no means points to starvation. In chewing, sufficient breakdown of plant material does occur to make available some carbohydrates, sugars and proteins for digestion and absorption prior to the fermentation process. This means that energy becomes accessible at once, rather than later.

Herbivore digestion: which type is best?

This question can more usefully be considered in evolutionary terms as a matter of functional adaptation to a certain set of environmental circumstances.

Although digestion in ruminants is more thorough than in non-ruminants, it is also much slower. The limiting factor here is the omasum valve which will not allow rumen contents into the true stomach and intestines for further digestion until fermentation is complete. The transit time of food from mouth to anus in the domestic cow is approximately 70 to 90 hours compared to only 48 hours in the horse. When good-quality food is readily available, this does not matter, but real problems can arise in climates with seasonal rainfall where grass contains progressively more cellulose and less protein during the dry months. Inadequate protein intake leads to loss of condition and weight. The ruminant cannot make up for this protein deficit by simply eating more, because the rumen has a fixed capacity and will not empty until fermentation is complete. If the protein content of the diet falls below six per cent, the animal will eventually starve. The non-ruminant has a distinct advantage under these circumstances, because it is able to compensate by simply eating more. There are no anatomical restrictions on its food intake and, despite a digestive process only two-thirds as efficient as the ruminant, the faster gut transit time and the simple expediency of eating twice as much will change that efficiency to four-thirds. Not having to spend endless hours ruminating also means that there is little need to invest in the development of complex digestive organs, which not only occupy much space but also weigh quite a lot. In the domestic cow, 40 per cent of body weight is taken up by the gut, compared to only 15 per cent in the horse.

Caecal fermentation evolved some 54 million years ago and was tried, tested and proven before the appearance of the ruminants some 15 million years later. Adoption of caecal fermentation required only minimal changes to the basic mammalian anatomy; the colon and caecum were already present and it was simply a matter of enlarging existing structures, not particularly demanding in evolutionary terms.

The evolution of the ruminant's stomach seems much more complicated in comparison. For a start, the first three chambers of the ruminant 'stomach' are in fact not derived from the stomach at all but are modifications of the lower end of the oesophagus. It is now believed that the initial expansion and enlargement of the foregut in the ancestors of ruminants had nothing to do with cellulose fermentation

The non-ruminant has a distinct advantage because it is able to compensate by simply eating more.

but functioned more as a structure devoted to nitrogen cycling and the detoxification of poisonous plant materials, both processes requiring the help of bacterial action. A modern example of the need to deal with plant poisons is provided by the koala bear, which feeds exclusively on the highly toxic leaves of the eucalyptus family. The koala has the ability to detoxify the harmful oils and is thus able to exploit a food source unavailable to other species. Prior to weaning, the mother koala feeds her own faeces to her offspring in order to prime its gut with the necessary bacteria. Enlargement of the foregut can therefore be seen as a form of preadaptation, already in existence and capable of further modification when the proto-ruminants eventually moved out of the forests on to the expanding grasslands.

It is now possible to see one of the reasons behind the mosaic evolutionary pattern of the horse. When the ancestral horse moved out on to the grasslands, modification of teeth was necessary to enable it to harvest a different food source, but no such demands were made on the mode of digestion, which correspondingly remained essentially unchanged.

The commitment to a vegetarian diet, and hence the need to maintain a prolonged, concerted attack on plant cell walls, forces all herbivores irrevocably into a certain way of life. Their waking hours are filled by the more or less continuous demand for food intake and the pattern of their lives is governed by the search for good grazing, often leading to massive seasonal migrations.

The grazing succession revisited

The different dietary requirements of ruminants and non-ruminants are addressed by the grazing succession of herbivores on the Serengeti plains (see also Chapter 1). Both the short and long rains falling on the Serengeti produce a profusion of fresh, protein-rich grasses. The zebra (non-ruminant), and wildebeest and Thomson's gazelle (both ruminants) can enjoy unrestricted access to the most nutritious grasses with little competition. As the grasslands dry out, this free-for-all is replaced by a system of differential grazing. The zebra can eat and digest the longer, coarser grasses with more stems and less leaf, which means more cellulose and less protein. As a non-ruminant, it is well able to cope with this diet, but is forced to eat a lot of it. The ruminant wildebeest, however, must seek out the shorter grass with its higher protein content. This herb layer first requires to be exposed, and the wildebeest relies partly on the zebra to do this, either by consuming or trampling the longer grass. Finally, the shortest and most protein-rich layer is exposed for the Thomson's gazelle. This energy-rich food is necessary for the smaller gazelle because it fulfils the needs of the higher metabolic rate that's characteristic of small mammals.

The meat eaters: beating indigestion

Having a diet consisting of high-quality protein not only obviates the need for cumbersome fermentation chambers, but also frees the animal from the drudgery of spending most of every day in the act of feeding. The breakdown of protein to peptides and amino acids by the enzyme pepsin in the stomach is a relatively straightforward matter, but is unnecessarily complicated by the manner of eating adopted by carnivores in general. A peculiar habit seems to be the need to gorge themselves with food, as though there were no possibility of another meal at any time in the future. Despite possessing carnassial teeth to slice meat into smaller portions, both cats and dogs gulp down large pieces of flesh until they are fully bloated. In this way, a grown lion may consume up to 50 kilograms of meat in a single sitting. A great white shark can dispose of a seal in a few bites and the large snakes such as pythons and anacondas, with their ability to dislocate their jaws, are able to swallow a goat as a single unit. With a mouth over a metre in width, the flesh-eating dinosaur Tyrannosaurus Rex must have been capable of swallowing huge chunks of meat, and of inflicting hideous injuries.

Carnivores are opportunistic feeders and can never guarantee success in making a kill or finding a carcass to scavenge on any particular day. The 'smash and grab' attitude to food undoubtedly reflects this inconsistency in food supply but this behaviour also demands the provision of a highly elastic stomach which can expand to accommodate large volumes of meat. In this regard the stomach functions partially as a storage organ. But a stomach bulging with large chunks of raw meat unfortunately creates problems. Digestive enzymes cannot hope to penetrate and breakdown this massive meat before putrefaction sets in – a process producing highly toxic and sometimes lethal chemical compounds. Nature has solved this particular problem by evolving a means to chemically sterilise the food and prevent bacterial growth. This is achieved by the secretion of a strong solution of hydrochloric acid derived from parietal cells lining the stomach wall. The benefit derived from this acid production is threefold: it kills bacteria which would otherwise cause putrefaction, demineralises bone so that calcium and phosphorus can be absorbed and produces an acid medium that greatly improves the efficiency of the enzyme pepsin in breaking down protein molecules. Hydrochloric acid is therefore a vitally important component of gastric juice, but its role is often understated. It is, of course, present as a necessary ingredient for protein digestion in the true stomach of herbivores and other vertebrates, but its function as a sterilising agent and a destroyer of bone confers on it a greater status in carnivores. However as any victim of indigestion knows, excess stomach acid causes unpleasant symptoms and may proceed to peptic ulceration or perforation of the stomach. Clearly the stomach needs protecting from its own digestive juices. In all groups, the secretion of copious amounts of mucous from the mucous glands is critical in preventing the stomach from digesting itself.

As with herbivores, the stomach contents of carnivores cannot pass into the further reaches of the alimentary tract until the digestive process has reached a certain stage. Even with a system of efficient protein breakdown, this period is likely to be prolonged simply as a result of the large size and sheer volume of the meat chunks consumed. Serial x-ray studies have tracked the initial digestive process in pythons, following their ingestion of rats. Within an hour, the rat can be seen to have reached the stomach of a two-metre-long snake. At 27 hours, the skeleton of the rat shows evidence of decalcification, and digestion is well under way, but it is not until almost 120 hours before the rat is fully 'dissolved' and the stomach contents discharged into the intestine.

For completeness, the stomach is also the source of one more essential substance – 'intrinsic factor', which is necessary for the

Both cats and dogs gulp down large pieces of flesh until they are fully bloated.

absorption of vitamin B12. Red blood cells cannot mature adequately without this vitamin, so a deficiency of intrinsic factor leads to pernicious (megaloblastic) anaemia, which can be fatal if left untreated. Before the ability of injectable synthetic vitamin B12, human patients were required to eat half a kilogram of raw liver each day in order to survive this disorder.

The digestive system in birds

One of the major requirements of the digestive tract in birds is that it should be as lightweight as possible without compromising the efficiency of the digestive process. Obviously, this is a necessity if the bird wishes to get off the ground. In some species, particularly vultures, when full to capacity with carrion, emergency lift-off in the presence of a perceived threat can only be achieved by vomiting a substantial proportion of the ingested food. A commitment to flight also brings with it dependence on a high-energy diet to subserve the demands of a high metabolic rate. Only the large, flightless birds are exempt from this imposition. Unlike vultures, most birds cannot gorge themselves with food, and the acquisition of a slow-burning fermentation system is clearly not an option. Birds therefore must devote a good deal of their time to the search for good-quality foods containing proportionally high levels of fat and/or protein.

Crops and gizzards

Most, but not all, birds possess a crop, an enlarged sac-like structure at the lower end of the oesophagus, which comes in a variety of shapes and sizes. With a number of different functions, it serves mainly as a temporary storage organ where seeds can be softened by water prior to digestion, or as a means of transporting food back to the young in the nest. But, curiously, the crop may act to produce food in its own right, or be utilised to form part of a defence strategy. 'Crop milk' is fed by pigeons and doves to their young. This secretion is formed from sloughed-off cells from the lining of the crop, which are rich in fat and protein. The whole mechanism is under the control of the hormone prolactin, secreted by the pituitary gland, and results in a fluid remarkably similar in composition to mammalian milk. And, as an effective means of defence, petrels can eject the foul-smelling contents of the crop at high speed into the face of an aggressor.

Peptic digestion occurs in the main body of the stomach, but further chemical breakdown requires the action of the gizzard, a pouch-like extension of the stomach with a thick, muscular coating. Strong muscular contraction in the gizzard serves to crush both seeds and bone and, to aid this process, many species swallow small stones or grit. In certain types of pigeon, the gizzard is equipped with a lining containing a number of hard, pointed projections, the better to crush tough seeds such as nutmegs. Indigestible materials – bone, teeth, claws and hair or feathers – collect in the gizzard of various species of raptor, to be regurgitated later as pellets.

One of the ways to maintain a short, lightweight digestive system is to improve efficiency of the process in a manner analogous to the ruminant herbivores, by allowing digestive juices a second chance of doing their work. This phenomenon is seen typically in warblers, whose dietary preference is for fruits with a very waxy exterior, and largely composed of fats. In these birds, the partially digested fruits leave the stomach to be mixed with enzymes, which are secreted into the duodenum and are capable of breaking down complex fats. The resulting mixture is then returned to the stomach where the vigorous muscular contractions of the gizzard churn the material to enable complete homogenisation to occur. High-energy fats can therefore be processed and utilised with substantially greater efficiency, without the need to invest in a longer intestinal system.

Dung – completing the cycle

Not surprisingly, the end product of all this eating and digestion is a mountain of dung. But it is Nature's rule that nothing is allowed to go to waste, including the waste products themselves. So dung becomes an important, integral part of the ecosystem, without which the cycle of life would be incomplete. It forms at once a source of food, a natural fertiliser and a means of seed dispersal.

The sacred dung beetle

The dung beetle, or scarab as it was once known, was revered in the culture of Ancient Egypt as the symbol of renewal and rebirth. Perhaps an indication of its importance in the feeding chain, there are approximately 7 000 species of dung beetle, varying in size from one millimetre to a giant six centimetres. They are found on every continent except Antarctica and form three main groups – the rollers, the tunnellers and the dwellers.

Male and female rollers meet at a dung pile where the male offers a giant brood ball, which may be up to 40 times his own body weight. If this gift is accepted, the pair roll the ball away, always using the back legs for propulsion, though the female may ride on top of the ball. The journey can be hazardous for the male, since it is likely that at this time he will be required to fight off the attention of would-be thieves. The ball may be rolled for distances of up to 70 metres from the dung pile before the pair bury it in a soft place. Mating then takes place and the male usually disappears to find new mates while the female divides up the dung ball into smaller spheres and lays an egg in each. She then seals them with a mixture of saliva and her own faeces and remains in attendance not only until the grubs emerge, but until they are mature enough for independence.

The specialised tunnellers bury their dung ball in harder ground, to a depth of some 30 to 40 centimetres, but depths of over a metre are not unknown. Hugh quantities of dung are buried, since it may need to sustain the larvae for periods of up to a year. Dwellers, on the other hand, contribute little to the movement of dung, being content to live, mate and presumably die in the parent dung pile. All told, about 75 to 85 per cent of dung deposited on the African plains is removed by the rollers and tunnellers. Placed at different levels within the soil, this provides a huge reservoir of fertiliser for the grasslands. Pits dug on the Serengeti by soil researchers showed 15 to 20 per cent of the soil to be made up of buried dung. And it is not just a matter of producing fertiliser, for, in its burrowing and tunnelling, the dung beetle also aerates and loosens the soil, providing a better environment for the growth of grass.

The indispensable elephant

Within the wetlands and the adjacent flood plains of the Okavango Delta in northwestern Botswana, numerous small islands are to be found dotted with ilala palms. The islands themselves have been formed by the activities of countless generations of termites, but the palm trees are immigrants, transported and delivered as seeds in the cargo holds of elephants, to commence their new life from a pile of elephant dung. Given the prodigious appetite of elephants, it is not surprising that a good quantity of seeds and fruits pass through their digestive systems. Their favourites include the seed pods of the apple-ring acacia and the fruits of the ilala palm and marula tree. Nature has determined that many types

The pair roll the ball away, always using the back legs for propulsion, though the female may ride on top of the ball.

of seed will germinate better if passed through the digestive tract of an animal. The elephant is not unique in this regard, and the help of other herbivores such as giraffe, buffalo, eland, impala and water-buck may also be enlisted. Some seeds with a tough external coat will *only* germinate after traversing an animal's gut, and yet others are so specialised that they will only germinate after passing through a specific animal. Thus the African eggplant and approximately 30 per cent of the larger tree species of West Africa are wholly reliant on the elephant for their reproduction. In this way, the elephant does not simply disseminate the seeds of a given species but also bears responsibility for the diversity of the habitat (Chapter 12).

A panoply of uses

A journey out on the African plains soon after sunrise is bound to reveal a troop of baboons sifting through piles of elephant droppings deposited during the night. The objects of their interest are undigested seeds, nuts and fragments of fruit. And what the baboons leave behind becomes fair game for a variety of seed-eating birds and squirrels. At the end of the dry season, clouds of butterflies descend on fresh dung to harvest moisture and, needless to say, dung is irresistible to all manner of flies, which deposit their eggs in the sure knowledge that emerging maggots will find a banquet suited to their particular requirements.

Remarkably, elephant dung also finds a place within traditional African medicines. It is used by women for a variety of gynaecological disorders, but particularly for uterine cramps associated with dysmenorrhoea or pelvic inflammatory disorders. The dung is burnt on a low fire and the woman squats over this in such a way that the smoke is able to enter her vagina. Inhalation of the smoke, on the other hand, is employed to relieve headaches, particularly migraines. How these therapeutic benefits work are unclear, though it seems that a volatile substance is released from the dung that induces relaxation of smooth muscle.

CHAPTER FIVE

NATURE RED IN TOOTH AND CLAW

predation and survival

... a whale sieving plankton from the sea ...

The imperatives of instinct

Life feeds on life, whether it's a fox devouring a rabbit, a whale sieving plankton from the sea or a vegan crunching a lettuce leaf. Despite the obvious truth of this, it is a notion that makes us uncomfortable at times. We have successfully sanitised our feeding habits, leaving the business of raw butchery to others, about whose activities we prefer to know little or nothing. Our social sophistication has allowed us successfully to disconnect the relationship between the steak and kidney pudding gracing our plate and the terror and violence of the abattoir. Most of us would be appalled to witness the dismemberment of an impala by a pack of African wild dogs, with an apparent savagery and speed almost incomprehensible to our sensitivities. But, shocking though the mechanics of this may be, the carnivore has no choice but to act on instinctive imperatives: it is simply responding to the most basic requirement for life.

Without predation, the herbivore population would expand to levels that could not be sustained by the environment, outstripping the available food supply with disastrous consequences, not only for the herbivores themselves, but also for the rest of the eco-system. We are witnessing something very much like this with certain elephant populations. Elephants, with no natural predators except lions from time to time, are confined, by and large, to national parks where their access to long-established ancestral seasonal migration routes is prevented by the incursion of human settlement and development on the fringes of the parks. The natural consequence of this is an assault by the elephants on the available vegetation, to the extent that forest is soon converted to savannah, with trees being pushed over in order to plunder the uppermost leaves, or being killed by having their bark stripped off. These activities alter the balance of the whole ecosystem and, to prevent conversion of the habitat to a lunar landscape, the artificial creation of park boundaries has necessitated an equally artificial solution, that of culling the elephant population (see Chapter 12).

In nature, a balance is established over very long periods of time between the numbers of predators and prey in a given location, and this is confirmed by the remarkably constant ratio between the population of lions and that of herbivores in East Africa. In Tanzania's Tarangire National Park, this ratio is one lion per 292 prey ungulates (hoofed animals); in the Ngorongoro Crater, one per 260; in Kagera Park, one per 300; and in Albert Park, one per 360. Early studies have estimated the average annual kill rate per lion to range between 20 and 36.5 prey animals, which would indicate that approximately 10 per cent of the herbivore population falls to predation by lions each year. Although this seems an awful lot, it is nevertheless a sustainable loss, given the prevailing birth rates.

Characteristics of predators and their tactics

Just as the herbivores share general characteristics, regardless of species, so do the carnivores. A successful predator is likely to have most of the following features:

- A dental arrangement that is capable of gripping, stabbing, slashing and slicing the flesh of its victims, which requires well developed canines and carnassial teeth.
- A mode of walking and running on the toes (digitigrade) like sprinters, rather than on the whole foot (plantigrade).
- Exceptional vision and hearing, which are the senses most used by the cat family: a dark, moonless night appears to a cat as a cloudy day would to us; and a lion can hear the tearing of grass by a herbivore a kilometre or more away.

- A body that is long and lithe, together with long legs, both adaptations for running at speed. This is further enhanced in the cheetah (the fastest land mammal) by a highly flexible spine and a long tail, which acts as a counterbalance.
- A coat designed to camouflage and conceal. The tawny coat of a lion resembles dry-season grass, and the broken stripes of the tiger are almost impossible to spot in the tall, waving grasses where it lurks to ambush or stalk prey.

Prey animals are often faster than their pursuers and, typically, in an attempt to escape at high speed, they underestimate the stamina of the predators and are eventually overtaken as they tire, sometimes at a distance of five kilometres or more. The strategies employed in hunting fall naturally into two main groups. The first of these is the long chase, and is typically found in the dog family – African wild dog and wolf – or 'dog-like' animals such as the spotted hyaena. During the chase, the emphasis is on endurance rather than speed, so that the prey animal is slowly brought to exhaustion. Killing is achieved by biting or tearing at vital structures but, in most cases, feeding begins while the victim is still alive. A different tactic is employed with small prey, which involves stalking and pouncing, then killing the victim by vigorous shaking.

Rather than involve themselves in a long chase, for which they have little stamina, cats prefer to hunt by stealth, using available cover to stalk and ambush prey, with only a short final rush. Curiously, the individual components of the hunt – stalking, killing and eating – appear to

The broken stripes of the tiger are almost impossible to spot in the tall, waving grasses.

function as independent units. This provides some sort of explanation for behaviour commonly seen in which lions, gorged to capacity, will stalk a passing prey animal; or why even hungry cats, having killed, may defer eating for several hours. Cats kill larger animals by strangulation, either by gripping the throat to crush the trachea, or by covering the whole muzzle with the mouth. Smaller prey is usually killed by a death-bite to the neck, in which a canine tooth is inserted between two of the cervical vertebrae to transect the spinal cord.

Straightforward theft by scavenging is another, less demanding, way of obtaining food and is widely practised by hyaenas, lions and jackals and, of course, by the vulture family, for whom it represents a full-time strategy.

Predators may hunt as solitary animals or in groups. Hunting as a pack obviously increases the chance of a successful outcome and explains the high kill/hunt ratio seen in African wild dogs, wolves and spotted hyaenas and, to a lesser extent, lions. On their own, carnivores will normally only attack prey that is smaller than themselves, but group hunting offers the opportunity of tackling larger animals. Thus a pride of lions can attack hippo, buffalo and elephant with reasonable expectations – an impossible situation for a single animal.

On the African savannah there is a definite hierarchy among predators, which is determined less by hunting success and more by the ability to defend ownership of a kill. All hierarchies are, to some extent, artificial and depend on the circumstances prevailing at any one time. For example, a cheetah may normally give up a kill readily enough to a hyaena, but if she has cubs to feed or is desperately hungry she will not hesitate to attack and drive off the would-be thief. Likewise, a solitary wild dog has little chance against one or more hyaenas, but a pack of wild dogs, by co-operating, will easily rout a number of hyaenas. Accepting these variations, the undisputed regent of the predators is the lion, followed in turn by the African wild dog, hyaena, leopard, cheetah and jackal.

The lion as hunter

The lion has been the most studied of all the predators, the subject of ongoing systematic scientific investigation on the Serengeti Plains in Tanzania for over 40 years, as well as in other countries such as South Africa, Namibia, Botswana, Zimbabwe, Kenya and Zambia. Being the top predator, the lion has a great variety of prey to choose from and it is not surprising that a medley of hunting techniques has developed to cater for this. A hunt for impala, therefore, proceeds along very different lines from a hunt for buffalo.

With buffalo, there may be little attempt at concealment and the pride may simply walk in the open towards a grazing herd. The idea here is to cause panic in the hope that buffalo will scatter in every possible direction, making it simpler to identify a suitable individual. How this choice is made is unknown and, although there may be a predilection to select young, old or slow animals, this is by no means always the case. The buffalo is a dangerous animal with formidable horns that can toss or stab any adversary and, as a consequence, the attack by lions is always concentrated at the rear end. The co-operation of pride members here is essential, since it may be necessary for one or two of them to act as decoys or worriers at the head end. A lion, or several lions in sequence, may attempt jumping on the back of the buffalo, either to cause it to collapse under their weight or, alternatively, to effect a bite to the spine that induces paralysis. Attacking from the rear puts the lion in a position to disembowel the victim using its sharp claws, or to bite in the region of the groin or testicles, both of which result in rapid blood loss. Once the buffalo is down, death is usually brought about by a throat hold. Other large prey animals such as the hippo and sub-adult elephant are usually attacked when they are alone. Notwithstanding the difficulties inherent in subduing and killing these formidable animals, some prides have made them a speciality food item. Large folds of fat preclude any possibility of a killing throat hold in either species, so death is brought about by causing huge blood loss or by inducing shock. There seems little doubt that the intense physiological and psychological stress induced by such an attack contributes a good deal to the onset of exhaustion – which renders the prey helpless.

While buffalo are large, fairly ponderous animals and not too difficult to overhaul in a chase, the pursuit of the more alert, swift and intelligent zebra requires much more in the way of stealth and cunning. In this situation, flanking movements are often employed, where lions move in crouching posture, using available cover to occupy positions not only along the flanks of the herd but also at points in the far distance, lying directly in the path of likely escape. To spring this trap requires one or two lions in the rear to break cover and initiate the stampede. Watching this action unfold, it is difficult to avoid the conclusion that each lion knows precisely what to do and is completely aware of its own role in the proceedings. Since no sound, or at least no audible sound, is exchanged between the lions, it seems to be a matter of responding to different postures and movements. In order to avoid powerful kicks from the zebra's back legs, which can kill or maim, the lion usually attacks from the flank. A paw is thrown over the shoulder or rump, the lion using its tremendous power to drag the prey sideways and backwards, a manoeuvre often accompanied by securing a bite hold on the back. Despite the almost military precision of the attack, the hunt often collapses in disarray when some of the lions are detected, make the wrong moves before the trap is sprung or when the herd escapes in an unexpected direction.

Smaller prey, such as impala, can be taken by single lions and this requires a slow, careful stalk followed by a short, running charge. Although lions can reach speeds of 60 kilometres per hour, they can maintain this for only 100 metres or so and therefore a successful stalk needs to take them within some 30 metres of their intended victim. Game of this size is usually caught by a slap on the haunch (not difficult with a paw the size of a dinner plate) or by tripping, and killed quickly by a bite to the neck or throat. As the dry season advances and surface water becomes restricted to a few pans, the waterhole becomes a favourite site for ambush. Lions conceal themselves in the surrounding grass or thickets, but even if the prey animals have detected their presence, they may still come to drink, driven, in desperation, by thirst.

Lions are acutely aware of the activities of vultures, and the sight of a descending vulture, though it may be a mere speck in the sky, will be enough to send lions off in the direction of a probable kill. A free meal has great appeal to lions and they will expropriate a kill from all other predators, scavenging, in truth, more frequently than the much maligned spotted hyaena.

Lions are territorial, and what they eat is determined by the local environment and the preferences of potential prey animals for particular habitats. Out on the open plains, seasonal herds of zebra, wildebeest, gazelle and tsessebe can be expected, while in more heavily wooded country, there is likely to be an abundance of smaller antelope such as impala and various small buck.

Lions are acutely aware of the activities of vultures.

Pieter Kat has been studying the lions in and around the Moremi Game Reserve in Botswana's Okavango Delta for the past five years. As part of this ongoing project, a detailed examination of the diet of two neighbouring prides, the Santawani and Mogogelo prides, has revealed some unexpected findings. Most studies of lions have shown that only three or four prey species make up 75 per cent of their diet, but in the Okavango study area, it took eight species to achieve this proportion and, in total, 19 prey animals were identified.

The Santawani lion pride lives in terrain where water is available only in seasonal pans and they prefer to hunt larger animals, usually in the hours of darkness. In contrast, the Mogogelo pride has a permanent water supply in its territory and their preference is to hunt smaller animals during the daylight hours, frequently by ambushing victims as they come to drink. In both cases, buffalo, zebra and wildebeest are available for only five months or so of the year, during the rainy season. This may go some way to explaining the eclectic nature of these lions' diets, since during the rest of the year they must make do as best they can. This does not explain, however, the surprising number of baboons eaten by the Mogogelo pride, a food item studiously ignored by the Santawani pride, despite its abundance. The answer to this most probably lies in the background of these particular animals. Lion cubs receive their hunting education by example and, depending on the availability of local prey items, may become highly skilled at killing certain species, for example giraffe, yet be quite unaware, as the Santawani pride appears to be, that baboons and warthogs are edible species. It seems likely that at least some of the core group of the

Mogogelo pride have an immediate ancestry that has taken baboons as a primary food source, while none of the Santawani pride has background experience of eating either baboon or warthog.

Despite the fact that prey animals have reasonably good night vision, lions are, by preference, nocturnal hunters. A stalking or motionless predator is more difficult to see in the dark and the relative coolness of the night surely gives the lion greater stamina in the chase compared to the blistering heat of midday. Contrary to common belief, a bright full moon is not a 'hunter's moon', and at these times lions can usually be found lounging around – an attitude that may change if the moon disappears behind a cloud.

Lions are intolerant of other predators and will kill leopard, wild dog and cheetah if they are presented with the opportunity to do so. Spotted hyaenas are tolerated at lion kills, but only at a distance, and they will be savagely attacked if they transgress acceptable limits. In turn, lion cubs or old and ill lions frequently fall to predation by hyaenas.

The success of a lion hunt varies dramatically with the number of lions involved, with a 17 to 19 per cent success rate for a solitary hunter versus 30 per cent for two or more lions. Despite this low success rate for individuals, hunting by a single lion accounts for 48 per cent of all lion hunts, with communal hunts involving more than two lions making up 32 per cent of hunts. Hunting is largely the prerogative of lionesses – some 85 to 90 per cent of the time – with the males trailing behind except when dealing with large prey such as buffalo.

Humans only rarely become food for lions, but the numbers are large enough to frighten tourists occupying flimsy tents. Camp fire horror stories are much embellished by tour guides, usually in an attempt to demonstrate their own intrepid qualities. Lions driven to eat human flesh are usually old, sick or hampered by dental problems – small consolation, one supposes, if a sudden tear appears in one's tent during the night.

The construction of the Mombassa-to-Uganda railroad between 1896 and 1901, in what was then British East Africa, provided the backdrop for the most celebrated episode in the history of man-eating lions. During a period of several months spanning 1897 and 1898, two large, maneless lions killed and devoured over 100 people, among them 28 Indian labourers brought into the country to build the railroad. The attacks took place in an area near the Tsavo River, where the lions

A bright full moon is not a 'hunter's moon', and at these times lions can usually be found lounging around.

became adept at taking men from their tents in the hours of darkness. Hysteria became so widespread that construction work was forced to a halt and workers deserted in their hundreds.

It fell to the lot of the chief engineer at Tsavo, John Patterson, to get rid of this menace and restore order. Patterson spent many nights sitting with his rifle in a tree over bait of tethered goats, but the lions always seemed to evade him and, at times, the hunt degenerated into farce or tragedy. On one occasion, fleeing from the lions, so many workers climbed a single tree that it broke, precipitating them back into the path of the man-eaters who, fortunately for the workers, had already left the scene.

Finally, after various failed attempts at entrapment, Patterson could write in his diary:

> 'At 8.15 pm the lion came up to carcass of donkey
> and I gave him a shot through the lungs and he
> died roaring and plunging in a terrific way.'

The dead lion measured 2.9 by 1.1 metres and was 'a most powerful beast in every way'. The second lion was finally dispatched 18 days later, to be followed by two days of riotous celebration. The den of these lions was eventually discovered and, in addition to bones, any number of bracelets and locally made trinkets were found – evidence that confirmed their man-eating habits had been well established prior to the arrival of the railroad crew.

The African wild dog

The African wild dog is the equivalent of the North American and European wolf and, like the wolf, it has been the subject of intense persecution by humans. As late as the 1960s, wild dogs were shot on sight by game wardens in the Serengeti and other national parks and are still shot by farmers whenever the opportunity arises. The causes for this animosity are difficult to unravel, especially since the dogs do not prey on domestic animals. Perhaps it's some psychological censorship on our part, reflecting an abhorrence of the methods employed by the dogs in securing a kill, which usually involves tearing the victim to pieces. Stories that wild dogs are responsible for game abandoning areas, however, are pure fiction and are not supported by direct observation.

Wild dog society revolves around a single pair of dominant animals (the alpha male and female) who have sole breeding rights, supported by a variable number – often 20 or more – of adults and sub-adults of both sexes. If a subordinate female becomes pregnant, her cubs are likely to be killed by the alpha female. Unlike lions, it is the male pups who remain with the natal pack, the females migrating at the age of about two years to a new pack of males. The home range of a pack may be huge, sometimes in excess of 1 000 square kilometres, depending on the density of prey, but the dogs are not truly territorial, there usually being considerable overlap in the ranges of neighbouring packs. By preference, they are nocturnal hunters, usually spending the day sleeping or resting, but with two definite peaks of activity in the early morning and evening. Their diet comprises mainly impala and other small antelope, with a seasonal glut of newborn zebra and wildebeest. They will, on occasion, tackle fully grown zebra.

When hunting large prey, a common tactic is for the dogs to advance in a front towards a herd to cause panic and identify slower or infirm animals, which they then pursue. This behaviour may be repeated several times before they isolate a candidate. Subduing and killing prey the size of a zebra presents a formidable challenge and can only be accomplished by a pack working together. Typically, a lead animal secures a hold on one leg while the others concentrate on disembowelling the prey. In some cases, a nose hold by one of the dogs immobilises the prey, in much the way that a professional horseman subdues a mustang. Hunting smaller prey, particularly impala, which make up approximately 85 per cent of their diet, requires an out-and-out chase. Sometimes an advance party flushes out prey from woody terrain, in the same way that game beaters flush out pheasants. The impala is certainly much faster over a short-distance sprint, but it lacks the stamina of the dogs who can maintain a speed of 56 kilometres per hour for several kilometres. In attempting to escape, the impala may zigzag or run in a large circle, which often allows trailing dogs the opportunity of cutting across the circle – behaviour that has undoubtedly led to the myth of dogs running in relay, with stragglers replacing tiring lead animals. This long-distance chase may last for five kilometres or so, but most prey are overcome in under three kilometres. When prey the size of impala or gazelle is captured, it is simply bowled over and disembowelled. A group of dogs pulling and tearing in different directions can reduce a carcass to just head, spine and skin within 10 minutes.

Spotted hyaenas are often to be found shadowing wild dog packs on the hunt, and will attempt to steal their prey once it is brought down and before the main body of the pack arrives on the scene. This sometimes works but, more often than not, the dogs will attack the hyaenas and drive them away.

Wild dog pups are born in an underground den in a litter of eight or more, spending three weeks in the rest chamber before emerging into the light. They are usually weaned by eight weeks. They are very vulnerable to predation at this age, and one of the ways to locate a wild dog den is to look for concentrations of vultures in nearby trees. After weaning, the pups engage in aggressive begging when the rest of the pack returns from a successful hunt – behaviour that appears to induce the adults to regurgitate food.

The hunting success of wild dogs is outstanding and far superior to any other carnivore. Although success rates as low as 39 per cent have been recorded, most studies indicate a rate of 70 to 85 per cent, which raises the question of why the African wild dog is the least common of African carnivores. Although the wild dog population in southern Africa seems to be relatively stable, this is not so in East Africa, where there has been a marked decline in numbers. The reasons for this are not clear, but human intolerance does not seem to be the major factor. Diseases such as canine distemper, rabies and others are likely to play an important role, originating, in all probability, from a reservoir of infection in domestic dogs living on the fringes of national parks. Canine distemper was responsible for the deaths of 1 000 lions in the Serengeti in 1994, illustrating the urgent necessity for creating buffer zones between parks and areas of human habitation. Although it has been popular to blame wildlife for outbreaks of disease in domestic animals, in reality the direction of transmission is almost always the other way around.

The misunderstood hyaena

Like the wild dog, the spotted hyaena wins no prizes for popularity, its oddly shaped, sloping back and furtive demeanour being sufficient to consign it to the realms of the unloved. An unlikely evolutionary product of the civet family – a nocturnal, racoon-like hunter – it is the commonest large carnivore in Africa, weighing up to 85 kilograms and possessing immensely strong jaws and teeth that can crush even the thickest bones. This ability not only to eat, but also to digest bone allows access to a food source denied other carnivores, which simply lack the highly muscular jaws to deal with large bones. The high intake of dietary calcium ensures that hyaena mothers can suckle their young for a period

Like the wild dog, the spotted hyaena wins no prizes for popularity.

of a year or longer. It also means that very little of a carcass is wasted. Unlike other carnivores, where up to 40 per cent of a carcass remains untouched, the hyaena will eat everything except the contents of the rumen and the horns of the larger antelopes. It has the ability, much like that of the owl but unique among mammalian carnivores, of being able to disgorge indigestible material in the form of a cast. A typical cast may include hair, horns, hoofs and grass, and sometimes stones, pieces of plastic buckets and other flotsam it may pick up in safari camps.

In areas where prey density is low, only solitary hyaena foragers may be present but, where large herbivores are common, hyaena society may consist of large groupings or clans numbering up to 60 animals, and occasionally more. The clans contain adults of both sexes, sub-adults and cubs, but are dominated by the females, including a matriarchal alpha female. Unlike wild dogs, any number of females may bear cubs but these are always of lesser rank than the offspring of the alpha female. Ongoing research on the Masai Mara in Kenya shows that, in the pecking order, all adult females are dominant over all males except for the sons of the alpha female. The cubs of the alpha female are privileged individuals who have high rank as their birthright, enabling them to eat better and grow larger faster. Although it has these hierarchical constraints, the clan does not function as a closely knit society in the way that, for example, a kinship group of closely related lionesses functions. There is no group suckling of young and, essentially, every hyaena must fend for itself in an openly competitive system. However, in places like Tanzania's Ngorongoro Crater, a collapsed volcanic caldera measuring some 16 by 8 miles, where herbivores are numerous but space is limited, the various hyaena clans operate as cohesive units and aggressively defend their territorial boundaries.

The spotted hyaena will eat any vertebrate that is, or has been, alive (it will eat carrion even though the body may be badly decomposed) and, like the lion, prefers the easy life, seizing any opportunity to steal fresh meat provided by the endeavours of lesser carnivores. It is frequently a solitary hunter and, since a single animal is capable of bringing down prey the size of an adult wildebeest, it does not strictly need to hunt in a pack. A chase initiated by an individual, however, is frequently witnessed by other hyaenas who are then likely to join in. But sometimes organised hunting as a pack does occur, particularly when going after larger prey such as zebra. The how, who and why

The how, who and why of deciding that zebra is to be on the menu on a particular day is unknown.

of deciding that zebra is to be on the menu on a particular day is unknown. The hunt can occur at any time of day, though usually not during the hottest hours. Like the wild dog, the hyaena embarks on a slow, loping run towards a concentration of game, with little attempt at concealment. Typically, a high-ranking female takes the lead, but more males than females usually participate in these co-operative ventures. The resulting flight of the prey group isolates stragglers, and these are run to ground in a chase covering a distance of up to two kilometres, a pursuit that relies on the stamina of the hyaena. Once caught, the victim is disembowelled. Thereafter, a feeding frenzy ensues in a desperate competitive scramble for food. The hyaena can eat up to a third of its body weight in a single sitting and a pack may completely consume an adult zebra in half an hour.

The intensely private life of the leopard

In contrast to the wild dog and hyaena, who are indifferent to their high visibility on the African plains, the leopard is a typical cat, preferring solitude and independence to camaraderie, and is disdainful of company. Despite being present in relatively large numbers, it is almost invisible, and the crowning glory of any tourist visit to Africa is surely to see a leopard. It seems to be at home in any environment where there is adequate cover for concealment and, like the urban fox, it does not shun human settlements. Indeed, the disappearance of many household pets, guard dogs and goats, is testimony to the ability of the leopard to adapt and prosper in more populated areas.

The private life of the leopard is precisely that – private – with its secretive nocturnal ways making it a difficult subject for systematic study. What is known is that solitary females inhabit defined territories within a larger male territory. With little contact between individuals, the various comings and goings of the day are marked out in scent trails, urine, dung and scratch marks on trees. These are matters of no small importance, since they may advertise the oestrus state of the female (see Chapter 10).

The diet of leopards comprises animal protein in almost any form, from beetles, rats and hares up to medium-sized antelopes such as impala. In fact, some 70 per cent of a leopard's diet comes in the form of small animals weighing 5 kilograms or less. In South Africa's Kruger National Park and in other parts of southern Africa, baboons also form part of the menu. An adult male baboon defending either itself or another member of the troop is a fearsome adversary, so that leopards hunt them by executing a hit-and-run manoeuvre in the darkness, when the baboons are sleeping in trees. For this reason, it is said that baboons prefer to sleep in knobthorn acacia trees, which have large numbers of rough protuberances in the bark, making them impossible for the leopard to climb. But this story, like many others from Africa, belongs in the realm of mythology.

The leopard is master of the concealed stalk-and-ambush, a tactic designed to take it within 10 metres of its intended victim in order to pounce before its prey can react in any significant way. If it misses, it seldom gives chase. After killing with the usual feline throat hold, leopards will frequently haul their prey, which may weigh as much as themselves, vertically up a nearby tree to cache in the branches, and feed intermittently over the next day or two. This pattern of behaviour seems to have developed as an adaptation to protecting food in areas where other predators, particularly lions and hyaenas, are common, because it does not occur in regions where the leopard is the sole predator.

The fastest animal on Earth

The cheetah's long, sinuous body is easily distinguished from the muscular, stocky outline of the leopard. The cheetah has characteristic facial tear marks and its spots are distributed individually, unlike the leopard, which has spots arranged in the form of clusters or rosettes. Although it still has a relatively large range throughout Africa, the cheetah is essentially an endangered species, with numbers in rapid decline. The reasons behind this include hunting, loss of habitat to human development, and competition from other predators. Being close to the bottom of the predator hierarchy, the cheetah can expect to lose 10 per cent of its kills to more dominant species. In addition, the loss of small cubs to lions and hyaenas makes it extremely difficult for cheetahs to survive in areas where the number of other predators is high.

In an attempt to mitigate against the threat of competitors, the cheetah has adapted to hunting during the daylight hours, a time when larger predators are likely to be sleeping or resting. Anatomically, with its long, sleek body and highly flexible spine, the cheetah is designed for the high speed chase and, as such, it preys on the faster antelopes, particularly Thomson's gazelle, springbok or impala. This is not an invariable rule, however, and it will just as easily take springhares or wildebeest yearlings. Relying as it does on speed for the most part, it prefers to hunt in more open country, using termite mounds and small hillocks as observation posts, and any low cover available to conceal a stalk. Although capable of speeds of up to 120 kilometres per hour, it can only maintain this for short distances of 300 metres or so and, like all cats, lacks the stamina for the long chase. With this limitation, a preliminary stalk, taking it to within 50 metres of the intended prey, is almost essential although, at times, a cheetah will deliberately trot in the open towards a group of grazing gazelle, seeming to select a victim at a distance of some 70 metres and then accelerating to an all-out sprint. Prey is brought down by a sideways swipe to the rump or hind leg, and a kill achieved by a throat hold. The dead animal is usually dragged into the cover of bushes, if possible, to avoid detection by vultures patrolling the sky who may then alert lions or hyaenas to

The cheetah uses termite mounds and small hillocks as observation posts.

the prospect of a free meal. The cheetah eats the muscles of the limbs, back and neck first, leaving behind the skin, digestive tract and most bones. Although every stalk does not result in a chase, 40 to 50 per cent of those that do are successful.

The female cheetah, like most cats, is by nature a solitary animal and any group found on the plains is likely to be either a coalition of males or a mother with her offspring. As with lions, it is the males of family groups that emigrate, travelling long distances to establish territories of their own. Male territories are, in contrast to those of leopards, incorporated into the much larger range of the female. These territories are fiercely defended and, while females may be granted access easily enough, any immature son travelling with his mother can expect a hard time. In Namibia it has been the practice of farmers and ranchers to shoot out large predators in order to protect their stock, but they are being encouraged to leave cheetahs alone, since they pose little threat to cattle. In these circumstances, a more gregarious social structure is starting to emerge, where groups of 10–14 cheetahs have been seen on a fairly regular basis. In the absence of competition, it appears that cheetahs not only seem to be more social but also co-operate at times in hunting, enabling them to tackle larger prey such as kudu or adult wildebeest.

The mortuary attendants

The Parsees are an ancient religious order in India, dating from the time of the Zoroastrians. In recognition of the sanctity of earth, fire and water, disposal of the dead by burial, cremation or casting the body into rivers is forbidden to them. Bodies are therefore placed in funerary sites open to the skies and elements. Colonies of vultures flock to these sites and are the acknowledged agents of disposal – they are venerated as carriers of the human soul.

Other than the performance of this ghoulish activity, vultures have an important role to fulfil in the natural world. Along with corpse beetles and other insects, they are the waste disposal service of the grasslands and savannah. This large, ugly bird with long neck and bald head, which spends its time hissing and jostling with its fellows over corpses, belongs to the group known collectively as griffons. Other members of the group include the white-backed and lappet-faced vultures and Ruppells' griffon. Surprisingly, these large birds may live for 40 to 50 years, mating for life and producing a single youngster

... the resulting assembly sometimes resembling aircraft in a holding pattern over a busy airport.

each year. Breeding is timed in such a way that eggs hatch at the end of the dry season – a time when the mortality among herbivores is at its peak and visibility over the ground is not hindered by green foliage.

Nesting sites in tall trees or on cliff faces are preferred. The paucity of trees on the true grasslands and the need to keep track of migrating herds commit the birds to travelling long distances in search of food. The energy expenditure involved in flapping flight is too high to make this a feasible option, and research involving Ruppells' griffons suggests this mode of flying would restrict the bird to a maximum foraging range of under 40 kilometres. Coping with return journeys of up to 300 kilometres clearly requires a different mechanism, and this is provided by soaring flight. The essential requirement here is a wing designed to provide maximum lift, achieved through a combination of large wingspan and slotting of feathers, particularly at the wing tips (see also Chapter 2). As the tropical sun warms up the morning air, it creates rising thermal currents, especially around prominences such as hills and cliffs. Vultures will launch themselves from their overnight perches only when the thermals are established and they utilise these as a means to gain height with minimal use of energy. The birds keenly observe the activities of others of their kind and, once one has found a suitable thermal, others will quickly follow, the resulting assembly sometimes resembling aircraft in a holding pattern over a busy airport. From great heights of a thousand feet or more, a large number of eyes are thus available to scan the ground for carcasses or the activities of predators and other scavengers.

Vultures are dedicated scavengers, eaters of carrion, and never kill their own food. The bulk of their diet is provided by the natural mortality occurring among animals, predominantly herbivores, rather than the leftovers of predator kills. The sight of a crowd of vultures engulfing a carcass, squabbling, pecking and hissing at each other, does not immediately suggest an atmosphere of co-operation, but co-operation of sorts there is. For instance, the often tough skin of a carcass needs to be breached in order to gain access to the internal organs, and this is most frequently achieved by the larger lappet-faced vulture with its strong beak. Thereafter, it's a competitive scramble for food. A long, hairless neck and skull are useful for feeding inside body cavities, since they prevent undue contamination by decomposing flesh. Smaller vultures, such as the hooded and Egyptian vultures, cannot compete with larger, more aggressive species and tend to lurk on the fringes of groups, snatching up morsels when they can.

Some degree of specialised eating is also apparent, with the white-backed vulture seeming to enjoy tough tendons and skin as well as soft parts. To supplement their diet, Egyptian vultures have also learned to break open ostrich eggs by dropping stones on them. The palm nut vulture is primarily vegetarian, feeding on the nuts of the oil palm tree.

For every plant eater, it seems there is a predator lurking somewhere. And every predator lives with the shadow of a scavenger or a decomposer close at hand. In Nature, there is nothing more rapacious than the selfish and relentless need for each and every organism to pursue its own destiny. It is the chain of life and it proceeds regardless of human approval or otherwise. It cannot be otherwise and, if it could, then it wouldn't be life.

CHAPTER SIX

WEBS OF LIFE AND DEATH

and the conservation of energy

They could not expect to survive long if they decided, on a whim, to pack their bags and head for the coast or the mountains.

Body over mind

'Death is not the inevitable end of life but only of the soma (body). Enlarge on this statement.' As an examination question, this would have been equally at home on a theology or philosophy paper; in fact it was part of a GCE 'A' level biology paper when I was a schoolboy. It was perhaps an unusually sophisticated way to test a student's knowledge, but it went to the heart of biology – the recycling of elements such as nitrogen, carbon and oxygen in the natural world, atoms that in a multitude of combinations form the molecules essential to the construction of living things.

The past chapters have looked at the origins of the grasslands and the unfolding of a sequence of events that have led to the creation of a living community. Within this community, each species is seen to occupy its own particular niche but is nevertheless dependent on other species for survival and wellbeing.

The origins of this system lie in the tectonic convulsions that thrust up mountain ranges many aeons ago. Within the rain shadows of these mountains, conditions emerged that were favourable for the invasion of grasses, which were consequently freed from their limited existence on the edges of forests. In time, creatures adapted to harvesting the grasses as a food source and converting it to flesh, and were then able to find a suitable home on the plains. In turn, these animals found themselves targeted by others who had a preference for meat rather than vegetation. The waste products of these animals, including their own corpses, returned nutrients to the land, out of which smaller creatures constructed their own intricate life cycles.

The web of life

In crude terms, we can represent this system as a food chain, but to do so would be to understate and simplify something of much greater elegance. The grasslands of the plains and savannahs should be viewed, more appropriately, as forming the centre of a web of life in which the grass itself forms the basic unit of currency. The whole system functions essentially as a single organism, where natural selection has populated these lands with animals and plants superbly adapted to the prevailing conditions. The wildebeest and the lion are not there by accident or chance but because they are fitted and equipped to be there. Moreover, their various characteristics more or less lock them into this particular niche, for they could not expect to survive long if they decided, on a whim, to pack their bags and head for the coast or the mountains. Putting aside the question of climate and its vagaries, the success of this community is largely defined by two things.

Obviously, the first priority is food, which is central to producing a society of mutual dependency. Any break in this food chain has consequences for those further down the line – the earlier the break, the more catastrophic for the whole community, with the potential for the whole system to collapse like a house of cards. Secondly, there is a need to establish a system of balance between the various residents, such that the numbers within individual species are never too high or too low to threaten the integrity of the whole community. These adjustments and balances have been worked out over periods measured in millennia and are highly sensitive to disruption. An epidemic caused by the canine distemper virus in 1993–1994, which resulted in the deaths of 35 per cent of the Serengeti lion population, had the potential to pose serious problems for the whole ecosystem. With less pressure from the top predator, the population of herbivores naturally enough increased at this time, but only, in turn, to present a danger to the grasslands themselves in the form of overgrazing. To a large extent, the consequences to the Serengeti were buffered by the sheer size of the grazing lands but, in a smaller area, the impact may well have been sufficient to decimate the grasslands, with repercussions for all its inhabitants.

The hub at the centre

Another way of looking at an ecosystem is to follow the trail of energy and materials that make up and bind the community together. The source of all life on Earth is ultimately the Sun, specifically the nuclear fusion of hydrogen to form helium, with the release of vast amounts of energy in the form of light and heat. Most of this solar energy is reflected back into space by the Earth and its atmosphere, with surprisingly little (0.1 per cent) being converted into usable energy in the form of products of photosynthesis. The basis of almost every food web is constituted by photosynthetic organisms, which together form the 'primary producers'. On land, these are represented by green plants and, in the ocean, by phytoplankton. Of the energy derived from photosynthesis, plants use 15 to 20 per cent to fuel the requirements of their own cellular respiration, while the rest goes into the production of new tissues. Approximately 10 per cent of the available plant material in the world is

One adult male lion is the
equivalent of a bale of grass
weighing approximately
140 tonnes — a very large
haystack indeed.

eaten by herbivores, the 'primary consumers', but only a fraction of the potential energy available can be truly assimilated to form flesh. Despite a digestive system adapted to breaking down cellulose and metabolising its products, much plant material passes through the gut of herbivores unchanged. What is assimilated by the animal goes largely to furnishing the needs of cell respiration, while only 10 per cent of the mass of vegetation consumed ultimately ends up as animal tissue.

In turn, a proportion of the primary consumers is eaten by carnivores, the 'secondary consumers'. The digestive tract of the carnivore is a much more efficient unit than that of its prey but, nevertheless, there is energy loss due to wastage and the ever-present demands of respiration. At the levels of efficiency actually measured by scientists, it is estimated that, to produce 15 grams of tissue, a lion needs to consume 1,500 grams of, for instance, wildebeest. And to create for itself this amount of flesh, the wildebeest would need to eat 10 kilograms of grass. Or, to put it another way, one adult male lion is the equivalent of a bale of grass weighing approximately 140 tonnes – a very large haystack indeed.

This tremendous drainage of energy, lost in the journey between the primary producers and the secondary consumers, is mitigated to some extent by a return to the soil of some nutrients in the form of dung and decomposing bodies. Organisms of decay, represented by various bacteria, fungi and invertebrates, are able to break down complex organic molecules, releasing simple compounds such as ammonia and carbon dioxide. These chemicals can then be recycled by plants to construct living tissue once again. Although this process does inject some energy back into the system, it does not begin to approach the levels necessary to balance the books. The failure to utilise all available materials, the inefficiencies of digestion, heat loss from warm bodies and the needs of cellular metabolism all contribute to a net loss of energy, which can be visualised as an ever-diminishing spiral travelling towards zero. And this is precisely where it would terminate, were it not for the availability of an ever renewable source of energy – the Sun.

Patterns of extinction

The K-T extinction (see Chapter 1) and its aftermath, together with the formation of mountain ranges, were precursors to the emergence of grasses as a new species, and their subsequent expansion to form one of the most important terrestrial ecosystems. These major events were not the result

Some believe the Sun to have a dark companion star.

of intent or design and are best characterised as accidental, random and unpredictable. Furthermore, events such as extraterrestrial collisions and changing conditions in the Earth's core and atmosphere become agents ready to deliver change in unpredictable fashion, where survival may become a matter governed, to some extent, by contingency – good luck rather than good genes. But over the past 50 years or so, tantalising evidence has appeared, suggesting that these random events might not be as random as previously thought.

In 1983 David Raup and John Sepkoski from the University of Chicago completed a detailed study on the extinction of a variety of marine organisms over the past 250 million years. Much to their surprise, analysis of the data demonstrated that patterns of extinction appeared to occur with a regular periodicity of approximately 26 million years. The match was not perfect, and their conclusions have since engendered a good deal of criticism and controversy, but they did lend some support to a suggestion by Alfred Fischer and Michael Arthur of Princetown University, in 1977, that extinction events during the past 250 million years occurred on a regular basis with a periodicity of 32 million years.

The causative factors responsible for such periodic mass extinctions are completely unknown, but this has not prevented widespread speculation. Cyclic events measured on a time scale of several million years are more likely to implicate extraterrestrial factors, and it is to this area that most attention has been paid. Some have suggested that the presence of a tenth planet, recently discovered beyond the orbit of Pluto (currently designated 2003 UB313) may divert the course of asteroids or comets on to a collision trajectory with Earth. More dramatically, some believe the Sun to have a dark companion star (again as yet undiscovered but already named Nemesis), which might also displace comets or asteroids from the Oort cloud, leading ultimately to terrestrial collision.

In the 1940s Jan Oort, a Dutch astronomer, proposed that 'long-period' comets with orbits measured in thousands of years had their origin in a large, more-or-less spherical disc around the Sun, which extended out from the orbit of Neptune to a distance approximately one third of the way to the nearest star. Asteroid bodies, he believed, remained in this distant 'cloud' until gravitational fields from larger celestial bodies caused them to change trajectory. It is now realised that the fate of most of these displaced comets is to be accelerated by gravitational encounters with the planets of the solar system, ejected from the system and lost forever. But what if certain planetary alignments, combined with other gravitational influences, could result in asteroid displacement in such a way as to make collision with the Earth a high probability? This might explain regularly occurring episodes of mass extinction, and it is rather discomforting to think that events set to shape the next such episode may already be in motion.

Be that as it may, there is certainly solid evidence that cyclical events do play an important role in climate, and hence evolutionary events, on Earth. These have been designated as the Lagrange/Croll/Milankovitch astronomical cycles, which have established that the Earth in its year-long journey around the Sun shows some change in its orbital pathway every 96 000 years. Similarly, the Earth's axis tilts from 21.5 degrees to 24.5 degrees and back every 42 000 years. And there is the further observation that the Earth wobbles as it travels through space much like a spinning top, each cycle taking approximately 21 000 years. Variations in orbit and axial tilt profoundly affect the amount and intensity of solar radiation reaching the Earth's surface and, as a consequence, induce periodic climate change on a global scale. These cycles have been responsible for numerous warm/cold stages, the expansion and retreat of the polar ice caps and, hence, variation in sea levels and the size of land masses. All these factors have been important at one time or another in the distribution and survival of most marine and terrestrial flora and fauna. The Earth is now entering another period of natural climate change, but the activities of our own species are likely to intensify this natural variation, with unforeseeable consequences.

It is clear that evolution is not driven by a single cause, but by a number of predictable and unpredictable events. Periods of mass extinction and astronomical cyclical events are but some of these phenomena, which serve, as it were, to 'reshuffle' the deck of life on Earth and, in the process, take life on to completely new pathways.

CHAPTER SEVEN

MULTIPLICATION GAMES
sex and reproduction

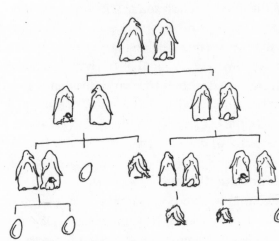

Nature has seen fit to impose a set of rules that determine who has the right to mate, with whom and when.

The reproductive imperative

If there is one thing that defines the difference between the world of the living and the non-living, it is the ability to reproduce – this is the essence of life itself. Reproduction represents the most powerful innate drive in mammals, once the more basic requirements of food, shelter and safety have been met. There is a compulsion to secure the future of the species and, at least in humans, perhaps to seek some measure of continuity of personal existence beyond the lifetime of the individual. In most mammals, rather than encourage a sexual free-for-all, Nature has seen fit to impose a set of rules that determine who has the right to mate, with whom and when.

Among some creatures, timing is of the essence. It would be catastrophic if, for example, the herbivores on the plains were to give birth in the midst of the dry season, when grazing may be poor and water only obtainable after a long trek. In these circumstances, birth is arranged to occur in the season of plenty, during the rains. This means, in turn, that mating must also take place in its own season. The predators, on the other hand, with some guarantee of an uninterrupted food supply, are to some extent freed from these constraints.

The question of who has the right to mate revolves, often, around a system in which a hierarchy of male dominance is established. What precisely constitutes dominance varies between species but, in general, it is determined by maturity, size, strength, fighting ability and health. Dominance may be expressed in a largely ritualistic display between competing males with little or no physical contact but, in some cases, it may take the form of a titanic struggle, leading to serious injury or the death of one of the participants. The prerogative of dominant animals to mate is Nature's way of ensuring that the genetic stock is not contaminated by the feeble and unhealthy, which may ultimately threaten the wellbeing and survival of the species. As it turns out, things do not always follow the plan, for often there are opportunists lurking, with an ever-ready eye for the main chance.

Spotted hyaenas: dominant females

It comes, then, as something of a surprise to learn that the most extreme example of sexual dominance involves not a male at all, but a female. Spotted hyaena society is dominated by females who outrank every male except for the sons of the alpha female. Males play no role in raising the young and are seldom tolerated close to the den. Early workers in the field believed that spotted hyaenas were hermaphrodites, possessing both male and female reproductive organs. This view was based on the observation that the female hyaena appears to have what closely resembles a fair-sized penis.

A detailed anatomical study, however, shows this to be a greatly enlarged clitoris. At the tip of the clitoris is an opening which extends backwards into a passage called the urogenital canal. This is formed from fusion of the urethra (the duct which, in females, is usually used exclusively for the excretion of urine) and vagina. This arrangement is unique among mammals, and its only counterpart occurs during a short period of embryonic development in the mole. The urogenital canal extends backwards for approximately 20 centimetres before giving off the urethra as a separate channel, which then joins the urinary bladder. In juveniles, the diameter of the canal is only slightly larger than that of the male urethra but, as puberty advances, it dilates progressively to approximately 50 millimetres, as needs it must to accommodate the male penis at a later stage. Unlike the penis, the clitoris does not contain erectile tissue, but it does contain a pair of powerful retractor muscles that run longitudinally along the length of the shaft. When these muscles are contracted prior to copulation, the effect is to shorten, widen and stiffen the urogenital canal, thus facilitating penile intromission. The same muscles are used to good effect during delivery of the foetus, allowing a shorter and less traumatic journey to the outside world.

This rather bizarre androgenisation (masculinisation) of the female external genitalia is the result of a peculiar profile of sex hormones. In the normal course of events, testosterone is a predominantly male hormone responsible for the development of secondary sexual characteristics (distribution of body hair, male physique, development of penis and testes), and is associated with more aggressive patterns of behaviour. Female mammals also produce small quantities of male sex hormones in the form of androgens, synthesised in the adrenal glands. In women, these hormones are concerned with muscle development and the distribution of pubic hair. What is unique in the spotted hyaena, however, is the high level of testosterone found in the blood of females, which at least equals, and often exceeds, that found in males. A study in the Masai Mara in Kenya showed the alpha female of one clan to have testosterone levels six times higher than the average male. This masculinisation appears to commence before birth, with blood levels of testosterone in female foetuses similar to those found in adults.

This particular hormone profile occurs exclusively in the spotted hyaena. It is not present in other members of the hyaena family, namely the striped hyaena and the brown hyaena, nor is it present in the likely progenitor of the hyaena, the civet, or indeed in any other mammalian species. How or why this arrangement came into being is essentially unknown, though it has been suggested that it represents an adaptation designed to promote better access to food in a society once dominated by aggressive males. While this may indeed be a consequence of the unusual hormone profile, it does not explain the absolute uniqueness of this phenomenon in the spotted hyaena – an adaptation that appears to have arisen completely *de novo*. A more likely explanation is the emergence of a random genetic mutation. Although most random mutations prove ultimately to disadvantage a species, some may be neutral in their effect and yet others may be beneficial. The spotted hyaena has not been disadvantaged by the production of testosterone in its females and, critically, its ability to reproduce has not been impaired.

The sex life of elephants

Unusual hormone patterns are also found periodically in male elephants. It has been known for many years that domesticated Asian male elephants exhibit recurring cycles of aggressive behaviour associated with an increase in secretions from temporal glands located just behind the eyes. Such behaviour is termed *musth*, a word derived from the Urdu expression meaning 'intoxicated'. Elephants in this state of mind can be calmed by providing access to females or by administration of a cocktail containing, among other things, marijuana, opium, camphor, rice and clarified butter. This phenomenon was thought to be unique to Asian elephants until the research of biologist Joyce Poole at Amboseli in Kenya proved otherwise. In a landmark study undertaken over several years in the 1970s and 1980s, she was able not only to document musth as an integral pattern of behaviour in bull African elephants, but also to characterise it in greater detail and to explain its physiological basis.

Male elephants usually leave the family group during adolescence. In fact, it is often their irritating and obnoxious behaviour that leads to their expulsion. Following this, they may wander alone, join another family group or associate in loose coalitions with other males. Occasionally, a group of youngsters may become attached to a 'patriarchal' male, with the group roving around in the ancient Greek fashion of the philosopher and his pupils. Bull units occupy distinct areas for months at a time, places that Poole described as 'retirement areas'. The hierarchy of dominance seems to be worked out at an early stage and depends largely on size. A typical threat or dominance display may consist of raising the head high such that the elephant appears to be sighting down 'the barrel' of his tusks. Ears are likely to be spread, and the head is often shaken vigorously to produce a loud cracking sound from the flapping ears. Alternatively, attempts may be made to mount (without erection) a subdominant animal. Submissive behaviour involves foot swinging, dust throwing, exaggerated feeding or placing the tip of the trunk in the mouth of the dominant animal.

Mature bulls are also sexually active in the absence of musth but, with the onset of musth, they leave the retirement area to spend a few days alone before actively seeking out oestrus females. Like the Asian elephants, their behaviour at this time becomes highly erratic and aggressive. Musth is a state of heightened sexuality, (perhaps the hormonal equivalent of Clark Kent stepping into the telephone box to emerge seconds later as Superman), in which a bull asserts his right to mate. It's a statement of dominance and a message to all that his time has come, regardless of his rank in other circumstances.

In addition to the profuse temporal gland secretion, Joyce Poole noted an almost continuous dribbling of urine associated with a distinct greenish discoloration of the penis. This latter was so marked that she initially entertained the possibility of its being related to some form of sexually transmitted disease. From the analysis of urine samples, collected before it soaked into the ground, she was able to demonstrate that the musth state coincided with very high levels of testosterone secretion. On the other hand, sexually inactive males showed very low levels of testosterone, while sexually active males who were not in musth showed intermediate values. Further observation of males indicated that periods of musth varied in duration from a few days only, in younger animals, to intervals of several weeks or months in mature males.

Female elephants are ready to mate at the age of 10 to 11 years, and advertise the onset of oestrus by odd patterns of behaviour and by emitting an oestrus call, which takes the form of an infrasonic rumble at high sound pressure, which may travel for several kilometres

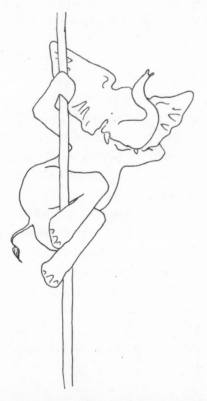

Female elephants are ready to mate at the age of 10 to 11 years, and advertise the onset of oestrus by odd patterns of behaviour ...

and which draws the attention of males (see Chapter 11). All sexually active, mature males have the right to breed, but the appearance of a musth male on the scene considerably alters the dynamics of the situation. The aggression of the musth male is such that other males avoid him if possible, which effectively routs the competition without a fight. Females seem to prefer mating with a musth male. He is, after all, displaying his manifest dominance. His presence also largely neutralises the threat of harassment from inferior suitors. In the Amboseli study, 65 per cent of observed copulations were performed by the largest 19 bulls out of a population of 160 males. And, within this group, musth males were dominant over all others.

Reproduction in the big cats

The hormonal high jinks of the hyaena and elephant do not occur in the cat family. But, being essentially solitary, cats have other problems, not least of which is to find a receptive mate in what may be a very large area. This process is aided by a form of chemical advertising (see Chapter 10). Oestrogens are female sex hormones, which prime the reproductive tract to a state of readiness for the subsequent stages of ovulation, fertilisation and the successful implantation of the fertilised ovum. A solitary female cat such as a leopard advertises this oestrus state through her oestrogen-rich urine and dung, which she sprays or deposits in prominent locations. The male is able to detect these increased oestrogen levels by smell and is thus able to find her by following the scent trail. Since this is likely to take time, the period of oestrus, of necessity, is required to last for some four to five days. But, given the often huge territories of the more solitary cats, the nearest male may be ouside of her range, or simply fail to find the female. In this case, she will drop out of oestrus and enter a quiescent phase until the onset of the next cycle – typically 46 days or so in the case of the leopard.

In humans and other primates the reproductive cycle is always accompanied by ovulation, regardless of whether that individual is sexually active or not. The cat family has evolved a different approach in which ovulation only occurs following copulation. This physiological adaptation is known as induced ovulation. It is not unique to cats, but also occurs in rabbits, ferrets, squirrels, mink, alpacas, llamas and other members of the camel family. For solitary animals with a low population density, there is a distinct advantage to this system in that the female of the species does not waste her ova or her energy on ovulation at times when pregnancy is not possible. Clearly, this approach has outgrown its usefulness in rabbits and other semi-domesticated groups, but they are stuck with an evolutionary mechanism put in place at a time when these animals were few in number and widely scattered.

Stimulation of the vagina produces a complex neurochemical response, which ultimately acts on the hypothalamus, a region of the brain intimately associated with the endocrine system and with emotional expression. The hypothalamus secretes a hormone (gonado-trophic releasing hormone) which acts on the pituitary gland at the base of the cerebral hemispheres. The pituitary gland, in its turn, secretes a further hormone (luteinising hormone), which induces maturation of ovarian follicles, leading to ovulation. The timing and frequency of copulation are of critical importance in induced ovulators. Experimental studies in domestic cats have shown that a single copulation during the oestrus period induced ovulation in only 20 per cent of females, while three copulations during a single day increased the rate of ovulation to 83 per cent; whereas a single mating usually fails to induce a luteinizing hormone response, serial copulation produces the sustained release of high levels of the hormone necessary for ovulation. In terms of timing, the probability of ovulation increases as oestrus progresses, so that mating on the first day has correspondingly less chance of resulting in pregnancy. In practical terms, this confers no great disadvantage to solitary cats, since it may take two or three days for the male to find the female. But clearly, when he does, the plan of action must be to mate with her as frequently as possible, since this offers the greatest guarantee of ovulation and subsequent fertilisation.

The special case of the sociable lion
Induced ovulation represents a good reproductive strategy for solitary cats such as the leopard, cheetah, serval and African wild cat, but does it function as efficiently in that most social of cats, the lion?

If reproductive success is defined as the ability to produce and nurture offspring to a stage where they are capable of an independent existence, then the success of lions can be described as mediocre at best. Circumstances often conspire to make this so and due consideration must be given to the high mortality rate among cubs, the suppression of ovulation following a pride takeover and the limited tenure of the pride male. Despite this, there can be no doubt that lions have successfully made the journey from a solitary state to a highly organised society. To arrive at some understanding as to why this was deemed to be essential and how it impacted on their reproductive success, it is necessary to examine some of the dynamics of lion society.

Earlier in their evolution, lions were more solitary animals and, as such, they inherited a reproductive process appropriate to this way of life. Current opinion tells us that they came together to form prides in order to occupy and defend a defined territory more easily, and to ensure a better defence for vulnerable cubs. Also, hunting as a group not only improved the chances of success but also provided access to a greater diversity of prey animals such as buffalo, hippopotamus and elephant, animals completely beyond the realms of possibility for an individual lion.

The core of a lion pride is composed of a group of usually closely related females who may number anywhere between two and 20. Actual numbers depend on the availability of prey species, and the average number of females is usually no more than six or seven. The territory of the pride is 'owned' by the females, and successive generations may have retained it for a number of years. Having vanquished the previous incumbent, the function of the pride male or males is to protect this territory from invasion, in exchange for which he is granted breeding rights.

This cosy relationship, however, is seldom as it seems. Pride structure is very fluid and it is rare to see all its members together on a regular basis. Females, either singly or in small groups, often go 'walkabout' and may be absent for several weeks at a time. Likewise, the pride male is frequently away on patrol. If the pride has been taken over by a strong male coalition, these males may deliberately set out to take over a second pride. A distinct fly in the reproductive ointment is the 'flightiness' of an oestrus female, who will accept any male during this period. If the pride male is absent, there is clearly a window of opportunity for any nomad in the vicinity to pass on his genes. And, as long as the pride male has copulated with this female at some time in the past, he will accept any subsequent cubs as his own.

The takeover of a pride by new incoming males is a time of great disruption, where aggression, fear and uncertainty permeate daily life, with ripples lasting for weeks or months. This can hardly be surprising, since one of the first actions of the new owners is to seek out and kill any cubs under the age of one year. Older offspring will also fall under the sword if they fail to make a swift enough escape. Abhorrent though this policy may seem, it does make some sort of biological sense. The incoming males are now in a position where they must defend the pride territory, sometimes at risk to their own lives. Why then should they take such risks to protect someone else's genetic investment? They have, after all, placed themselves in danger not just to acquire territory but to pass on their own genetic inheritance. Lionesses do not passively acquiesce to the murder of their young and may defend them with fierce aggression. In the event of a takeover by a coalition, however, it is a battle that they invariably lose. But they do nevertheless have the option of leaving the pride, taking their youngsters with them.

When life has settled down somewhat under the new regime, there is usually a round of vigorous copulation, where females appear to solicit mating from any or all of the incoming males. Researchers have advanced a number of theories to account for this behaviour. The most convincing proposal is that the death of their cubs prompts females to come into early oestrus. This view is supported by the observation that, in normal circumstances, the female will not mate until her cubs are approximately 18 months old. The death of these cubs is thus likely to lead to an early reintroduction of her oestrus cycle.

Whatever the truth of the matter, it has been clearly established that these early sexual skirmishes do not result in pregnancy. This suggests that females in some way possess the ability to manipulate their own reproductive physiology in order to avoid pregnancy. Since this is simply not the case, it is necessary to look elsewhere for the reasons behind this period of infertility.

As mentioned, the takeover of a pride by new males causes considerable upheaval and stress to the females, particularly so if their cubs are killed. Humans are also familiar with the effects of stress and are well aware that its ramifications spread far beyond the obvious feelings of anguish. The hypothalamus, the brain structure concerned with emotional matters and the workings of the endocrine system, is intimately involved in formulating the response of the body to stressful circumstances. It initiates and co-ordinates a variety of actions that include: (a) activation of the sympathetic nervous system, which primes the animal for fight or flight; (b) an increase in the coagulability of blood (an evolutionary adaptation designed to stem the flow of blood in the event of physical assault); (c) suppression of the immune system (which means infections of various sorts are more likely in times of stress), and (d) alterations in hormone balance. From the reproductive standpoint it is the change in hormone balance that is critical. It is more than likely that, during the stress of a takeover, the hypothalamus shuts down the secretion of gonadotrophic-releasing hormone. As a consequence of this, the pituitary gland is unable to produce luteinising hormone, and hence maturation of ovarian follicles and ovulation are not possible. A period of infertility after a takeover is therefore inevitable. Oestrus, meanwhile, can continue since this involves a separate metabolic pathway. Women are also well aware of these matters, since chronic stress may be associated with an interruption of menstruation or irregularities in the menstrual cycle. Research on the Serengeti has shown that it takes, on average, a period of 134 days for a lioness to become pregnant following a takeover.

The makers of wildlife documentary films seem unable to avoid showing the audience snapshots of tender moments between a lioness and her cubs, usually to the accompaniment of soft, melodic music. Unfortunately, the reality is that lions are rather poor mothers. Cubs can be left behind as the pride moves to a different location. They are frequently left unattended when their mothers are hunting and easily fall to predation by hyaenas at this time. Cubs also have no special priority at kills and must fight for morsels as best they can amidst the jumble of snarling, aggressive adults. The combination of maternal

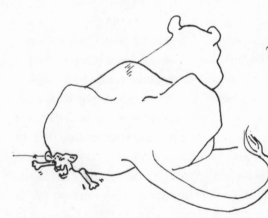

The reality is that lions are rather poor mothers.

neglect, starvation, infanticide and disease ensures that only 50 per cent of cubs survive to their first birthday. In the Serengeti, 67 per cent of cubs failed to survive to the age of two years in the 1960s, a figure that increased to 86 per cent over the next decade.

Given that a lioness avoids further mating following the birth of cubs, often aggressively so, for a period of up to 18 months despite the re-establishment of normal oestrus cycles, and since the average tenure of the male is only two years (and rarely extends to four years), the cumulative periods of infertility, pregnancy and postpartum reluctance determine that he may be restricted to only one bite of the reproductive cherry, at least with this particular pride. Which, of course, does not prevent him from peddling his wares elsewhere with another pride or as an opportunistic nomad.

Simple arithmetic shows that, following a pride takeover, the incoming male cannot expect to see his own cubs enter the world for at least 8 months (an infertility period of four and a half months plus gestation of three and a half months). This means that if reproduction is delayed or tenure is terminated before two years, then the cubs will be of an age when they are highly vulnerable to infanticide by the new males. The net reproductive output may thus essentially be zero, which represents an enormous wastage of life. Clearly, a critical component of reproductive success is the tenure of incoming males. This explains in large measure the increased success of a coalition of males compared to a single individual. A coalition has a far greater chance of repulsing a takeover bid and has the added advantage of sharing out the mating duties. A prolonged tenure also allows the possibility of fathering a second group of cubs.

One other aspect of lion reproduction that needs to be considered relates to the very high frequency of copulation which takes place between an oestrus female and her consort – a feat unmatched anywhere else in the animal kingdom and presenting something of an enigma. During a four-day period of oestrus, the honeymoon pair may mate every 15 to 30 minutes. It has been calculated that approximately 3 000 matings go into producing a single cub. Even allowing for the need for frequent copulation to induce ovulation, lion behaviour nevertheless seems rather extravagant. The rate of production of spermatozoa in the testes cannot hope to keep pace with these demands – to do so would require nuclear-powered testicles. Matings from the second day onward are therefore likely to be sterile. So why the need for a four-day marathon?

The only way a male can ensure that any future cubs will be his is to remain with the female throughout the entire oestrus period. Given that during this time she will accept any male, including opportunistic nomads, this unusual mating behaviour may represent an adaptation designed to prevent the lioness from wandering away in search of other partners, and thus contaminating the genetic succession.

Sexual dimorphism

In physical appearance, men and women are clearly different. Similarly, on the African plains, few people would fail to distinguish between a fully grown male lion and a lioness, the shaggy mane together with a greater bulk and muscularity being sufficient to identify the male. But in many species these differences in appearance, technically termed sexual dimorphism, are not so obvious. Apart from the presence of a bag of testicles, there is little to distinguish between the male and female in such diverse species as domestic dogs, cats, wildebeest, zebra, buffalo and many others. The obvious question to ask here is why it should be necessary for some species to develop and parade secondary sexual characteristics while in others it appears to be quite unnecessary.

In the search for answers in biological science, it is useful to think in terms of how any given adaptation serves the immediate needs of a species, or how it may confer an evolutionary advantage. An example to illustrate how these principles may work in explaining the contrast in appearance between males and females is provided by a specific feature of the antelope family – their horns. Despite wide variations in their social organisation, male dominance is expressed by body size and the degree of development of the horns.

The case of the kudu and the dik-dik

The greater kudu is a large, non-territorial antelope belonging to the bushbuck tribe and inhabits woodland and thicket. Females, which do not possess horns, form small groups of usually no more than four to five individuals who maintain lasting social bonds. The males are blessed with the largest horns in Africa, measuring up to 180 centimetres in length, with an average of 120 centimetres. Male kudu form loose-knit bachelor groups at times and only associate with females during the breeding season. Males are polygamous, which means they attempt to mate with as many females as possible. To stake a successful claim for exclusive access to a group of females, the male must vanquish other males in a system of open competition. The arrival on the scene of a mature male with large horns and an impressive body size is enough to lead to the withdrawal of smaller suitors, but more aggressive behaviour occurs if two equally matched males meet, though physical combat is rare.

This behaviour is in marked contrast to that of the dik-dik, a dwarf antelope of woodland glades and savannah, measuring no more than 40 centimetres or so in height. Both sexes possess short, backward-pointing horns. Social organisation centres around the formation of monogamous couples who defend a defined territory against intruders and who avoid 'extramarital liaisons'.

There are clear distinctions between the reproductive behaviour of these two species. In the polygamous, competitive system adopted by the kudu, the right to mate is restricted to a dominant male who is required to display and demonstrate his dominance to counter the claims of other potential suitors. And his dominance is expressed both by general body size and the degree of development of his horns. Any physical or behavioural trait possessed by a male that gives him an advantage over his peers is likely to confer on him a greater chance of reproductive success. Differences in physical appearance between males and females are, therefore, an indication of the rivalry existing between males for the right to reproduce. And, the greater the rivalry, the greater will be the difference in appearance. This phenomenon is also common in birds such as the peacock, where extravagant plumage in the male contrasts sharply with the dowdiness of the female.

In a monogamous species such as the dik-dik, the necessity for highly developed secondary sexual characteristics is made redundant. A high degree of sexual dimorphism in this setting would serve little purpose, since the male does not have to compete with, or deflect, the attentions of other males and therefore has no need for elaborate demonstrations of dominance.

The care of offspring

It seems rather strange, when so much effort and ingenuity go into the business of reproduction, that the care of offspring can be so variable in quality. In some species parental care may be a very desultory affair, while others engage in no parental care whatsoever, abandoning their young to the world at large even before they are born. The urgency to

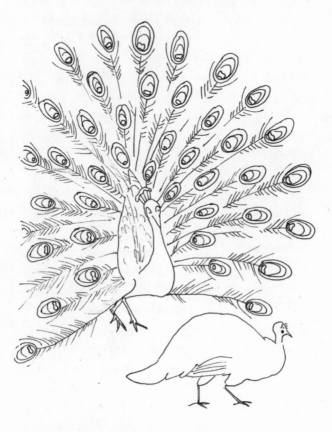

The extravagant plumage in the male peacock contrasts sharply with the dowdiness of the female.

ensure genetic continuity seems in many to have missed out a vital step. And this parental indifference goes some way to explaining, in part, why some species compensate for these deficiencies by producing large numbers of offspring so that even in the face of heavy predation, some will survive unscathed. But in some species such as the ostrich, care of the young is taken to remarkable lengths.

Ostrich ingenuity

The ostrich is a large, rather bizarre, flightless bird resembling in many features the now extinct moa of New Zealand, and preferring to live in arid regions of semi-desert and savannah. The eggs of this huge bird are the largest in the avian kingdom; each egg yields more than a dozen good-sized omelettes. The San bushmen of the Kalahari, in addition to enjoying the high-quality protein the eggs provide, use the shells as water-storage vessels.

Ostrich society functions on the principle of dominance, where a single male consorts with a harem of females. But within the female group there is a 'major hen' which dominates the rest of her sex. All the females lay their eggs in one nest and these are incubated only by the dominant male and major hen – though incubation is perhaps a misleading word here, since the eggs are largely being protected from the desert heat rather than being kept warm. The incubating pair can, however, accommodate only 20 or so eggs, and any excess eggs are turfed out of the nest to perish. But the major hen does not eject eggs at random – only those of lesser females. Brian Bertram, who studies these birds in Kenya, discovered that the major hen has the ability to recognise her own eggs. By numbering each egg as it was laid, Bertram knew precisely which eggs belonged to which females. Even if he shuffled the eggs around such that those of the major hen were on the periphery (and hence likely to be pushed out), she promptly moved them back into the centre of the brood. He concluded that individual eggs can be recognised from the pattern of pores on the surface – the air holes which allow oxygen to penetrate the shell and which are necessary for survival of the embryo.

If this is remarkable, what follows is more so. Once the chicks are hatched and reasonably mobile, their parents set out to capture and kidnap, by force if necessary, the brood of another pair. This behaviour may continue until a large number of chicks has been accumulated, or until a more dominant pair steals the whole lot for themselves, and is thought by many to represent a means of protecting offspring from predation. Young chicks are highly vulnerable to the attentions of all carnivores but, if the offspring of a particular pair can be hidden within a large crèche, the probability of at least some of them surviving is somewhat increased.

In these vignettes of animal reproduction it becomes apparent that sex has as many rules as a board game. But the rules are in place because reproduction is the agency through which natural selection operates. Reproduction is not essential to the life of a particular individual but it is critical to the survival of the species as a whole. Under these circumstances it is hardly surprising that Nature imposes a self-regulating system that's designed to sustain this agenda with the greatest possible efficiency. The quirks and tricks employed in achieving this goal represent the adaptations of a species to its own peculiar circumstances, and are a testament to Nature's ingenuity.

CHAPTER EIGHT

DICING WITH DEATH

monkeys, lions and AIDS

With FIV, the main route of infection appears to be through penetrating bite wounds.

First rumblings

In the late 1970s physicians on both seaboards of the United States and later in West Africa became aware of an unusual clustering of certain medical problems. Patients presented with a variety of infections, some of which were only rarely seen in medical practice, including a particularly virulent form of pneumonia caused by a protozoal agent, and a type of meningitis caused by a fungus. These conditions were often associated with an equally rare form of tumour – Kaposi's sarcoma. The common denominator linking these features was a profound depression of the immune system, involving wholesale destruction of lymphocytes – cells that determine the response of the body to invasion by infectious organisms or the presence of foreign tissue such as transplanted organs. Because of the primary effect on the immune system, this new disease was designated 'acquired immunodeficiency syndrome' (AIDS).

A feline parallel: FIV

In a remarkable parallel, a similar condition was discovered in domestic cats by veterinary surgeons in the mid 1980s. Here again, the clinical presentation was usually with a variety of infections associated with a marked depression of the immune system. Subsequent research demonstrated a viral origin for both these diseases. Although structurally distinct from each other, both viruses were found to belong to a group called retroviruses (or lentiviruses). This type of virus was not unknown and one had been previously identified as the causative agent in a rare type of human leukaemia.

Rather than DNA (deoxyribonucleic acid), the core genetic material in retroviruses is represented by RNA (ribonucleic acid). It is only when a virus infects a living cell that viral DNA is synthesised from RNA, a reversal of the normal process (hence the term 'retro'). This operation requires the action of an enzyme – reverse transcriptase. The DNA so produced then becomes spliced into the genome of the host cell, thereby 'coercing' it to produce more viral DNA and protein.

The virus isolated from cats became known as the 'feline immunodeficiency virus' (FIV) and that from humans as the 'human immunodeficiency virus' (HIV). It soon became apparent, however, that there were two types of HIV. The one we loosely call HIV is properly known as HIV-1, a highly lethal virus with a worldwide distribution including central, eastern and southern Africa. HIV-2, on the other hand, is much less virulent and is largely restricted to West Africa. Both variants of HIV are transmitted sexually and by direct blood-to-blood contact – including from blood transfusion and contaminated needles. There is 'vertical' transmission from mother to child, which occurs at the time of delivery and not via the placenta. Although sexual and vertical transmission both occur with FIV, the main route of infection appears to be through penetrating bite wounds. Despite the ability of HIV to replicate itself every six minutes, it usually takes several years before failure of the immune system becomes clinically apparent. But when it does so, survival is limited to two years or less in the absence of antiviral medication.

The African green monkey: SIV

Not surprisingly, the discovery of these viruses and their lethal potential has raised some crucial questions: where did the viruses come from, and are they new or are they mutations of existing organisms? Studies have been extended to a variety of other animal species to determine the distribution and biological effects of the viruses. And, importantly, attempts are being made to identify the likely ancestor of the HIV virus and establish if and how transmission could occur from one species to another. Research has now clearly demonstrated the presence of HIV-related organisms in a number of mammals including horses, sheep, cattle, goats, cats and several Old World monkeys. The work of Max Essex, Phyllis Kanki and others showed a very high rate of infection, between 30 and 70 per cent, in African green monkeys, and they were subsequently able to isolate the virus responsible, which is now known as the simian immunodeficiency virus (SIV).

The African green monkey, probably better known as the vervet monkey, has a widespread distribution throughout most of subSaharan Africa except for the Congo Basin and the Namib and Kalahari deserts. It is a small creature measuring some 60 centimetres in height and weighing less than 10 kilograms, and is well known to campers and safari operators for its habit of plundering any available food store or rubbish area. There is an even closer association between vervets and humans in West Africa, where the monkey forms both an integral part of the diet and an item of trade. Under these circumstances there is clear potential for cross infection to humans in the form of bites and scratches inflicted during capture, together with direct blood contact during preparation of meat for cooking.

The blood of HIV-positive prostitutes in West Africa, as would be expected, contains antibodies to proteins found in the core of the virus and its surrounding envelope. What is remarkable is that these antibodies are indistinguishable from those found in the blood of SIV-positive African green monkeys. This strongly suggests that the virus has been successful in jumping the species gap between monkey and human and is a possible contender for the role of precursor to HIV. Although HIV infection in West Africa produces clinical AIDS, the virus in this location is much less aggressive than its counterpart in the rest of the world. So what happened to this virus to make it so lethal? Similar comparisons of blood between AIDS patients in the USA and SIV positive vervet monkeys show that, as with West African patients, the antibodies to the core viral proteins are identical; but in the case of the envelope proteins there is no such correspondence. Clearly, somewhere along the line the West African virus (HIV-2) had mutated to a more virulent form (HIV-1).

Robert Gallo, who shared the Nobel Prize with Luc Montagnier for the discovery of HIV, believes that the progenitor of HIV originated in Africa, where humans and other primates were infected, and subsequently took passage to the Americas during the years of the slave trade where it eventually mutated to the virus (HIV-1) that we know today.

There are, however, good reasons to doubt the truth of this particular scenario. AIDS in Africa is and always has been a disease of heterosexual contact, and thus affects both sexes. Assuming that heterosexual relations, whether forced or otherwise, occurred at an early stage between the victims of the slave trade and their white 'masters', the distribution of any emerging AIDS-like disease would still be expected to be predominantly heterosexual. Whereas, in fact, when AIDS first appeared in the USA it was essentially limited to the homosexual community – only later coming to involve both sexes, a phenomenon no doubt aided and abetted by the use of contaminated blood and blood products in medical practice, together with the sharing of contaminated needles among intravenous drug users. These observations raise the possibility that other vectors may have contributed to the dissemination of the virus (see Fatal vaccinations? on page 103).

FIV – a benign affliction?

The appearance of an AIDS-like disease in domestic cats and the isolation of FIV-led research workers to cast a wide net in an attempt to identify similar viruses in populations of wild felines. In 1989 Margaret Barr at Cornell University discovered FIV in several captive exotic cats as well as wild panthers resident in Florida. Stephen O'Brien at the National Cancer Institute in Maryland found FIV in a large percentage of African lions. In eastern and southern Africa, it appeared that 80 to 90 per cent of lions were infected, with only the lion populations of Namibia, the Mana Pools region of Zimbabwe and the Umfolozi Game Reserve in South Africa being apparently free of infection – regions largely protected by hostile geographical boundaries. To date, FIV has been found to infect lions, cheetahs, pumas and bobcats, but not tigers or jaguars.

The high incidence of infection seen in lions indicates that FIV is endemic in this species. Both sexes are equally affected. It has been difficult to prove that vertical transmission occurs from mothers to cubs at the time of birth or during breastfeeding, because so few young cubs have been tested. What does seem certain is that most lions are

FIV-positive by their fourth birthday. Not surprisingly, scientists were very concerned about these findings – if FIV proved to be as lethal in wild felines as it was in domestic cats, then many species might be threatened with extinction.

Over several years Stephen O'Brien has made a detailed study of the gene sequences of FIV in lions. In this way it has been shown that FIV exists as several variants, each of which is largely restricted to a particular locale. The virus found in the Serengeti is thus slightly different from the one encountered in South Africa. By examining hundreds of blood samples from various cat species around the world, O'Brien came to believe that FIV was an extremely old virus, with origins going back hundreds and perhaps thousands of years. In his view, FIV first appeared in the original ancestral cat species, before this diverged into the many species we see today. This, he believed, explained the wide geographical distribution of the virus. He also put forward the elegant idea that, because of this very long exposure to the virus, there came to exist a form of co-evolution between virus and host in such a way that expression of a full-blown immunodeficiency syndrome had been lost through a process of natural selection. Despite its high prevalence, FIV was to be regarded as essentially benign in wild felines. It was only when the virus infected a similar but unrelated species – such as transmission from a bobcat to a domestic cat – that an AIDS-like disease would be possible, because each type of FIV was species specific. If this theory is correct, it provides reassurance regarding the future of lions and other exotic cats. But is it correct?

Elegant though this theory may be, it does leave many awkward questions unanswered. If the virus really arose in the earliest cat ancestor, why has it left tigers and jaguars untouched? And, especially with the very high prevalence in lions, how has the entire population of lions in Namibia and other regions managed to escape infection? These inconsistencies would seem to indicate that FIV appeared on the scene much later than has been suggested.

The apparent lack of pathogenicity (development of disease) stands on even more fragile ground and is based on a single study by Craig Packer and his associates from the University of Minnesota, who have been studying the lions of the Serengeti for many years. FIV is endemic in this population, with over 90 per cent of lions infected. With some obvious reservations, Packer concluded that FIV did not increase mortality among infected lions. But, in fact, it is not possible to draw this conclusion in a group where the rate of infection is so high – simply because no adequate control group exists as a basis for comparison. It would be the equivalent in medical practice of investigating the influence of smoking on mortality by studying a group composed of 95 per cent smokers – the statistical analysis would be clearly meaningless without a matched group of non-smokers.

Likewise, the inference that FIV is benign, based on genetic sequencing of the virus, does not stand up to close scrutiny because not one of the studies paid more than cursory attention to what was happening in the immune system itself. If FIV behaves in a way comparable to HIV, then the immune system would remain intact for several years – it is only when viral replication reaches a critical level that failure of the immune mechanism becomes inevitable. In these circumstances, a single blood sample taken during the early course of the disease may show absolutely no abnormality of immune function, but this data cannot be extrapolated to represent the state of affairs several years later.

The long-term future of cats

The first major cracks in the supposedly benign nature of FIV started to appear when Alessandro Poli described in detail the death of an FIV-infected lion at the Pistoia zoo in Italy. In the final weeks of its life, the lion exhibited all the clinical features typically found in the end stages of the immunodeficiency syndrome observed in domestic cats. In a more comprehensive and extended study undertaken over a seven-year period,

Most lions are
FIV-positive by
their fourth birthday.

Marta Bull and her colleagues at North Carolina State University set out to determine whether FIV infection in captive lions was associated with destructive changes in the immune system similar to those observed in domestic cats. In a direct comparison between matched groups of FIV-positive and FIV-negative lions, they were able to demonstrate a progressive failure of the immune system in all the FIV-positive animals. Moreover, this immune system failure was accompanied by the development of clinical illnesses typically linked with immunosuppression – illnesses so severe as to require euthanasia of all affected animals.

Despite the unequivocal nature of these results, some scientists – perhaps fearing the implications – have been reluctant to accept the findings on the basis that captive lions are more vulnerable to a range of infections not normally encountered in the wild. This may be true, but it does not detract from the fundamental observation of the failure of the immune system induced by FIV, which then opens the doors for a whole raft of potentially fatal opportunistic infections – in a way identical to the effect of HIV in humans.

In an ironic twist to the story, researcher Melody Roelke, working in Stephen O'Brien's laboratory, has recently published a study which, for the first time, shows abnormalities in the immune system of wild African lions infected with FIV. The changes observed precisely mirrored those found in humans with AIDS. Here is proof that failure of the immune system is not an isolated quirk restricted to populations of captive lions but is something which threatens to engulf the whole species.

As it happens, an opportunity presented itself that might unravel the question of whether FIV has an adverse effect on morbidity and mortality in free-ranging wild lions. Throughout the 1990s, over 90 per cent of lions resident in the southern part of Kruger National Park died. The immediate cause of death was bovine tuberculosis, contracted by eating the flesh of infected buffalo. The buffalo had, in turn, been infected by domestic cattle living on the fringes of the park. The Kruger lions have a very high prevalence of FIV (over 90 per cent), so the obvious question to ask was whether FIV infection made the lions more vulnerable to tuberculosis. The question is made more compelling because there is an exact parallel in humans where, at least in Africa, there is a strong association between HIV infection and tuberculosis. In sub-Saharan Africa, tuberculosis represents the single most serious complication of HIV infection, occurring in 50 to 70 per cent of AIDS patients.

Preliminary, and as yet unpublished, research has shown mycobacterium bovis – the organism that causes tuberculosis in cattle – to be a highly lethal pathogen in lions, irrespective of their FIV status. What is clear, however, is that lions harbouring FIV are more susceptible to tuberculous infection and die quicker, just like their human counterparts infected with HIV. And these observations may provide a clue in helping to explain another massive die-off in a lion population. An epidemic of canine distemper virus killed approximately 1 000 lions – one third of the population – in the Serengeti over a two-year period between 1993 and 1994. As in the case of tuberculosis in Kruger, the Tanzanian lions were infected by invasion of the National Park by diseased domestic animals – in this case dogs. Is it possible in this case also that the high mortality occurred, in part, as a direct consequence of failure of the immune system in a population of lions with a high prevalence of FIV? The answer is unknown at the present time but is now the subject of ongoing research.

The available evidence points clearly to the conclusion that FIV is not a benign virus in exotic cats but behaves in a manner analogous to HIV, causing, over time, a progressive failure in the immune system, rendering the animals more susceptible to potentially lethal opportunistic infections. On this basis, it would be a great mistake to be smug or complacent about the long-term future of these cats. And this problem is made more acute by the realisation that the main source of killer infections has proven to be domestic animals on the edge of the National Parks.

One obvious solution is to create a physical barrier between livestock and wild animals. But this will bring a great responsibility to park authorities because, left to their own devices within the fixed perimeters of the park, the inmates – particularly elephants – have the capacity to destroy their own habitat if population growth exceeds the carrying capacity of the land to feed them. In practical terms, this means that active management of animal populations cannot be avoided if the diversity of the environment is to remain intact.

Fatal vaccinations?

Poliomyelitis vaccine is a safe, well tested and established medical tool, which has undoubtedly saved many lives and prevented disability in thousands of young people. Its role in poliomyelitis prevention is beyond dispute but, in the mid 1970s, the indication for its use

The inmates
— particularly
elephants — have the
capacity to destroy
their own habitat.

was extended to include treatment or prevention of herpes, a sexually transmitted disease. This required the use of a live vaccine and the culture medium used to ensure viral replication was made from the kidney cells of the African green monkey. In 1976 the US Bureau of Biologics found that a number of samples of this vaccine, in addition to containing numerous polio viruses as expected, also contained particles later identified as type-C retroviruses. As a consequence, the vaccine was withheld from the market for a period of 20 months, while the importance or otherwise of these contaminating particles was debated. Nevertheless, the vaccine was eventually released. The argument developed by the manufacturers was that, since this was an oral vaccine, any contaminant virus could not enter the body through the intact lining of the intestinal tract. This much is true but what about the use of the vaccine in a group of homosexual men undergoing treatment for herpes? The herpetic sores themselves and the possibility of rectal and anal tears in this social group largely rendered this argument invalid. Here was the provision for entry of retroviruses through damaged membranes.

The obvious question flowing from this information was, therefore, whether the AIDS epidemic initially involving young homosexual men in the USA had its origins in the use of contaminated poliomyelitis vaccine prepared from viruses cultured in cellular material from the African green monkey – a species well known now (but not in the 1970s) to harbour SIV. In scientific circles these possibilities have been largely dismissed as speculation, but they have still not been adequately answered.

The potential for accidental transmission of disease from animal sources during therapeutic procedures does not stop at poliomyelitis vaccine. Organ transplantation is now well established in medical practice, but the supply of donor organs is substantially less than the demand and currently only one in three patients placed on a waiting list will receive a transplant. The direct corollary of this situation is that many will die before treatment can be offered. The domestic pig is an animal with organs such as the heart, liver, pancreas and kidneys very close in size to those of humans. But the direct transplantation of organs across species (xenotransplantation) is associated with a very rapid and aggressive rejection response.

The initial successful work in cloning pigs is now being extended to produce a race of 'knockout' pigs, so that pig organs transplanted into humans will not generate this aggressive early rejection. This involves inactivation of the gene for alpha-1,3 galactosyltransferase, an enzyme synthesised by the pig, which produces substances recognised by the human immune system as belonging to foreign tissue. Another approach is currently under investigation by breeding transgenic pigs, in which a fragment of the human genome is spliced into the pig DNA in the hope that this will completely eliminate rejection problems.

If this work comes to a successful conclusion, how certain can we be that it will not result in the transmission of hitherto unrecognised viruses completely alien to humans, with consequences impossible to predict? And this scenario may be compounded by the deliberate suppression of the immune system in the recipients of such organs by medicines now in use to control rejection of transplanted organs. Modern medical and biological research, particularly in the field of genetics, tests the boundaries of ethics and morality as never before. Because we can do something does not necessarily mean we should.

CHAPTER NINE

STRATEGIES FOR SURVIVAL
coping with extremes

It is a time when hot winds scour the Earth, and every water hole is reduced to a mosaic of cracks and fissures.

Surviving heat

There are times in Africa when the thing you wish for most is to turn back the sun, to press that great chromatic disc back below the horizon for a few more hours. In southern Africa the months of October and November are known, with some justification, as the 'suicide months', where the heat builds and builds until it seems the atmosphere itself will burst into flames. Following long months without rain or a cloud in the sky, it is a time when hot winds scour the Earth, throwing up huge tornado-like dust devils that march across the landscape. The air vibrates in the heat, vegetation wilts, grass becomes dry and brittle and every water hole is reduced to a mosaic of cracks and fissures in a barren bed of clay. With shade temperatures reaching 40°C on a regular basis, the land looks and feels exhausted, buckling under the weight of oppression and inducing, at least in some human residents, a peculiar mixture of lethargy and apathy combined with an ill-defined restlessness and anxiety.

Since the rains will not appear until December or January it is not surprising this is a time of great stress for the grass eaters. The paucity of both food and water determines that most will have to migrate to areas close to a permanent water source and those who fail to do so are likely to provide bounty for predators and scavengers. But some – Grant's gazelle, the springbok, gemsbok and eland – choose to stay. Despite the absence of surface water, they are able to obtain sufficient moisture from their food; this consists, in the main, of the leaves of shrubs and trees possessing long root systems able to reach the water table, together with the underground water storage organs – bulbs and tubers – of other plants and supplemented with fruits such as wild melons and cucumbers. But regardless of these particular adaptations, all animals, whether browsers, grazers, predators or reptiles, are confronted on a daily basis with the problem of dealing with a massive heat load from the environment.

Life in the desert

The Namib is one of the world's oldest deserts, stretching in an unbroken line along the southwest coast of Africa from the Skeleton Coast in the north, a region littered with the bones of shipwrecked vessels (both ancient and modern), to the Orange River in the south. Extending up to 100 kilometres inland, this world of featureless gravel plains, rocky outcrops and towering, shifting sand dunes receives less than 25 millimetres of annual rainfall. When the cold, deep Benguela Current from Antarctica hits the continental shelf off the West African coast, a great upsurge of cold water occurs. The surface temperature of the sea falls, as a consequence, to a level that completely inhibits the formation of rain-bearing clouds. But, as if in compensation, this same current that denies rain creates instead dense coastal fogs that may drift 50 kilometres or more inland on sea breezes. And, without the fog, no life could exist here at all. The great age of the Namib has, unlike other deserts, allowed the evolution of a rich variety of organisms, many of which are exceptional; but all of them are required to meet the challenge of daytime surface temperatures of up to 70°C, freezing nights and the ever-present prospect of desiccation. And the most vulnerable organisms in these extreme conditions are the plants.

Desert plant strategies

In more temperate climates plants are kept cool by the evaporation of water through small pores – stomata – located on the underside of their leaves, water that has been drawn up in a continuous column from underground roots. At the same time, the open stomata allow entry of carbon dioxide, which is essential for photosynthesis (see Chapter 1). With a readily available and reliable source of water, this process of evaporative cooling (transpiration) is highly efficient. But it does require a lot of water – an option not possible in the desert where water must be conserved at all costs.

Described originally in 1852 by a German botanist, Frederick Welwitsch, the plant that bears his name – Welwitschia mirabilis – is unique to Namibia. Although it is really a tree, it prefers to hug the desert surface, spreading its large, tattered-looking succulent leaves (of which it only ever produces two) along the ground and only seldom exceeding a height of more than a metre during a life span thought sometimes to exceed 2 000 years. Like members of the cactus family, it opens its stomata only at night. Absorbed carbon dioxide becomes chemically fixed and is then stored overnight in large cell vacuoles. At sunrise carbon dioxide is released into tissue spaces where it becomes available to participate in photosynthesis – now behind closed stomata. This chemical variation has been termed CAM photosynthesis (crassulacean acid metabolism) after its discovery in Crassulacean plants. It is a feature found in orchids, lilies, grape vines, asters, pineapples and daisies – all of which are drought-resistant species.

Strategies of small desert creatures

The huge dunes of the southern Namib may give the impression that no life could take hold in this beautiful yet barren place. But this apparent emptiness is an illusion because beneath the ocean of sand exist surprising numbers of small creatures that literally swim in the sand, using it as a subterranean highway, shelter, larder or home. From the desert surface downwards, the temperature falls progressively and at a depth of one metre it remains within the range of 12 –18°C. Small desert animals exploit this by burrowing down to avoid the heat. Their policy is one of evasion rather than endurance. A typical day will therefore find them underground during the hot hours, with periods of activity on the surface mainly restricted to the cool of the night.

Obtaining life-giving water in a land of no rainfall tests the ingenuity of Nature to its limits. On foggy days dune beetles emerge from the sand to work their way to the crest of the dunes. Here they face into the fog-bearing breeze, lowering their head so that droplets of water condensing on their body will run down towards their mouth. In this way they can obtain enough water to increase body weight by an impressive 35 per cent.

The north American desert burrowing cockroach has evolved a somewhat different approach. Protruding from its mouth are two small pouches (hypopharyngeal bladders) covered with closely spaced fine hairs of no more than 160–180 nanometres in diameter. When relative humidity exceeds 83 per cent, water vapour condenses on the hairs, causing them to swell with accumulated water. Absorption of water into the body occurs when the hypopharyngeal bladders are flushed with 'saliva' (in reality an ultrafiltrate of endolymph), which creates an osmotic gradient, 'dragging' water from the swollen hairs. This cycle can then be repeated indefinitely.

Reptiles, such as the sidewinding snake and various lizards, can obtain fog water directly by licking condensation from their body scales. But for most others, including spiders, scorpions and small mammals, water is delivered second hand from the bodies of their prey. In this way, the ephemeral vapour of fog becomes the focal point of an intricate web of life – just as the most basic currency of life on the savannah is grass, without which neither herbivore nor predator could exist.

But if obtaining water is difficult enough then retaining it is just as much of a problem. Having a small body may carry the advantage of being able to escape the heat by burrowing, but it creates difficulties in regulating body temperature and water balance. A small body has a high surface area relative to its volume and therefore has a strong tendency to gain or lose heat and moisture rapidly – presenting the unenviable prospect of death by incineration during the day or freezing at night. So it is not surprising that adaptations designed to prevent desiccation and cope with wide temperature fluctuations are particularly well developed in the smallest of desert inhabitants.

Insects are furnished with a hard, shiny external skeleton that reflects heat and minimises water loss. Nevertheless, some water loss is inevitable and occurs through the breathing tubes (spiracles), which open directly onto the body surface for the uptake of oxygen and elimination of carbon dioxide. Most of the water lost by desert insects occurs in this way and may account for as much as 70 per cent of their total water loss. One way to combat this is to close the spiracles for a time and simply stop breathing, an action that necessitates complete cessation of all physical activity in order to reduce metabolic rate and carbon dioxide production. In some insects ventilation occurs differentially in such a way that air intake occurs only through the spiracles at the head end of the body while expiration is

Some reptiles can obtain fog water directly by licking condensation from their body scales.

confined to the rear. This unusual arrangement lends itself to operating a condensation system to limit water loss. Such is the case with tenebrionid dune beetles, which possess an air cavity (subelytral cavity) below their wing cases, protecting the spiracles in this region. Thanks to the highly reflective exoskeleton, the air in the subelytral cavity is somewhat cooler and permits water vapour from expired air to condense and run down the body, to be reabsorbed by the anus.

The response to heat of birds and mammals

The body temperature of birds and mammals is controlled within very narrow limits, varying between 35 and 40°C, depending on species. Why evolution has 'chosen' this particular range is not absolutely clear, but it is recognised that enzyme function is optimal within this range and declines sharply at lower or higher temperatures. Having a body temperature higher than the ambient environmental temperature also means that excess heat can be removed by convection and radiation. But under the conditions found in the desert or African savannah, where air temperatures can exceed 45°C even in the shade, animals and birds are confronted with the problem of massive heat load. Behavioural changes typically seen in small creatures are not applicable to larger animals, which must necessarily develop more sophisticated adaptations to their physiology.

Most of the organs in the mammalian body can tolerate a limited rise in temperature for short periods without ill effect, the exception being the brain and, curiously enough, the testicles. Anyone who has had a bad cold is aware of the effect of fever on brain function: it becomes difficult to maintain concentration and we become forgetful and lose some co-ordination in performing delicate tasks. Higher fever brings outright confusion and hallucinations, which may progress to coma and death. A body temperature of 43°C is usually fatal. Meanwhile at the other end of the body, the production of spermatozoa is so sensitive to temperature that even normal body temperature may inhibit it. As a consequence testicles are located outside the body cavity in protective scrotal sacs where the local environment is cooler.

The purpose of fever is still a matter of hot debate among scientists, but it is recognised that the effect of fever is not always detrimental. An increase in body temperature, for instance, is known to kill a number of invading bacteria, irrespective of other responses generated by the immune system, and it is argued that such an adaptation may have evolved as the original purpose of fever. Certainly in the pre-antibiotic era this observation was exploited in medical practice to treat a variety of infections, including syphilis. Syphilis is a venereal infection caused by a spirochaete – treponema pallidum – which, according to tradition, was brought back from the New World to Europe by Columbus' sailors. Known colloquially as 'morbo gallico' (the French disease) it was endemic in northern Italy by the 16th century and spread rapidly all over Europe. Although readily treatable now with penicillin and other antibiotics, one of the many interesting previous treatments was deliberately to induce a state of high fever by artificially infecting patients with the malaria parasite. The malarial fever in many cases was sufficient to kill the syphilis spirochaete and cure the disease. Of course this sometimes left the patient to contend with recurrent bouts of malaria, but this was judged to be preferable to the prospect of insanity or sudden premature death.

Large mammals have evolved a number of strategies to combat the dangers of excessive heat. The simplest is to present to the outside world a light-coloured, shiny coat that reflects as much solar radiation as possible. Trapped within this coat is a layer of still air; a relatively poor conductor of heat, air can function as an insulator in cold weather or as a heat barrier in hot conditions. The overall effect depends on the density and thickness of the coat and, in a tropical setting, it needs to be short (less than 20 millimetres in length), dense and flat. It also helps if a continuous layer of fat beneath the skin can be avoided so that the body is better able to function as a radiator. In desert-adapted animals, body fat may be localised to one particular spot – in the camel and in zebu cattle this is in the hump, while in desert sheep it is located at the base of the tail.

Cooling off without dehydrating

Evaporation of water from any surface, animate or inanimate, causes cooling. And just as evaporation of water from leaves keeps plants cool, so sweating and panting cools animals' bodies. In some insects and reptiles the source of this water is their own urine or saliva spread over the body surface, but in mammals it is usually cutaneous sweat glands. Cooling the skin surface also cools blood circulating in tiny capillaries in the subcutaneous tissues. And as this blood travels back into the general circulation it helps to maintain core body temperature within acceptable limits. In many species, particularly those with long noses – the bovids, horse family and the giraffe – evaporative cooling occurs mainly via the membranes lining the nasal cavities and upper respiratory tract.

Panting through the nose, combined with an increased rate of breathing, causes rapid evaporation of moisture from nasal membranes – and hence cooling. Like the skin, the nasal membranes have a rich blood supply that can carry cool blood back to deep body structures. Panting also carries the advantage of retaining salt, a substance lost in substantial quantities during sweating. Faced with the same hot surroundings, a dog, which pants, will need to replace only the water it has lost, whereas a human will need to replace salt, in addition, if major electrolyte imbalance is to be avoided.

Evaporative cooling provides an efficient means of regulating body temperature during exposure to heat, but its success depends on ready access to a reliable source of water. Given the seasonal nature of rainfall on the savannah, some species are forced into a cyclical migration. Water forms some 70 per cent of the weight of mammalian bodies, and depletion of water at a cellular level causes metabolic chaos. On a macro level, extracellular fluid compartments shrink, blood volume diminishes and blood viscosity increases to the consistency of treacle. Death becomes increasingly likely as a result of heart failure or heat stroke. A person undertaking manual labour under tropical skies may lose three to four litres of sweat per hour and can expect to be dead when 10–15 per cent of body weight has been lost in this way. The threat of dehydration confronts every creature that finds a home in a hot environment; and, critical to any mechanism designed to combat heat is the equally important question of precisely how much water loss can be tolerated without serious consequences. As with insects, the larger animals inhabiting the savannah have come up with some ingenious solutions.

Elimination without waste

In all species the metabolic breakdown of complex molecules such as proteins produces a number of waste products that need to be excreted, either because they are toxic or because they cannot be recycled. The final product needing to be eliminated from the body varies between species but is determined in large measure by their dependency on water. With an infinite volume of water at their disposal, fish excrete ammonia – a highly toxic and lethal substance for most other forms of animal life but which, in fish, can be so rapidly diluted that adverse effects are avoided. Mammals have taken matters one stage further and chemically combine ammonia with carbon dioxide to form urea, a much less toxic material. Urea is excreted in urine and, although urine can be concentrated to limit water loss, nevertheless a certain minimal volume of water loss is inevitable. Unless, of course, you can recycle the urea.

The gut of all herbivores contains bacteria that chemically break down the tough cellulose walls of plant cells, allowing the animals' digestive juices to get to work on the protein-rich cytoplasm (see Chapter 4). In exchange for this service, herbivores 'donate' urea to the intestinal bacteria for use in synthesising bacterial proteins. Recycling of urea in this way may have evolved primarily as a means to provide herbivores with a supplementary protein source during times of dietary deficiency. But it has proved to be of benefit as a method of water conservation at times of heat stress and limited availability of water.

Conservation of water is of even greater importance to birds, where a high metabolic rate places them at greater risk of dehydration. Here the excretory pathway from ammonia to urea takes one further step to produce uric acid as the final product – a substance insoluble in water and therefore requiring an absolute minimum of moisture for its excretion. This method is also favoured by insects, land snails and reptiles.

The case of the gemsbok

Despite the almost complete absence of rainfall, some species such as Grant's gazelles, eland and gemsbok (oryx) appear quite at home in the arid and semi-arid regions of Africa. The gemsbok seems to represent the ideal symbol of endurance, standing alone or in small groups in the oppressive heat of the featureless wastelands of Namibia and the Kalahari thirstland. It may stand for hours without shade under a fiery sun and, perhaps as a consequence, possesses a volatile temperament, being quick to engage in aggression, often using its one-metre-long, rapier-like horns with devastating effect.

With adequate water supplies the gemsbok, like the eland, prefers to cool itself by sweating. But, in the face of water shortage, these animals have adapted the more dangerous policy of passively allowing their body temperature to rise – a method that works on the principle that heat always flows down a thermal gradient from a warmer object to a cooler one. If the animal allows its body temperature to rise to match that of the environment then progressively less heat is absorbed.

If a gemsbok is deprived of water under experimental conditions and subjected to a simulated desert environment, its first response is to cease sweating. As air temperature climbs above 39°C, the animal allows body temperature to rise, maintaining it at 0.5–2°C above ambient levels, even when air temperature reaches 45°C. Lacking this adaptation, a wildebeest or buffalo faced with similar conditions would continue to sweat and ultimately would succumb to severe dehydration and heat stroke. Under natural desert conditions, the heat accumulated by the gemsbok during the hot daylight hours is dissipated by radiation during the cooler night hours. But to carry off this trick successfully, the gemsbok requires a means of protecting the brain from potentially lethal increases in body temperature.

Evaporation of water from the linings of nasal passages in long-nosed animals, such as the gemsbok, cools the blood passing through these structures. Inside the skull this cooler blood drains into a large blood-filled space known as the 'cavernous sinus'. The carotid artery carrying hot blood from the core of the body passes through the cavernous sinus on its way to the brain but, instead of doing so as a single channel, it divides into several smaller branches – exposing a much larger surface area of arterial blood to the cool venous blood. The system functions as a heat exchanger, ensuring that arterial blood going on to supply the brain is delivered at a safe temperature even though core body temperature may be several degrees higher. This anatomical assembly – known as the 'rete mirabile' ('wonderful network') – appears to function equally well during vigorous exertion when even more body heat is being generated by exercising muscles. The carotid rete is not unique to desert mammals and is found in all types of antelope, bovids, sheep, camels, giraffe and many others. In the giraffe it may also play an important role in protecting the brain against large fluctuations in blood pressure as the animal stoops to drink.

The myth of the camel

Despite its reputation for having an evil temper, the dromedary camel has been domesticated as a beast of burden throughout Arabia and Africa's northern deserts for centuries. Its amazing ability to go without water for long periods of time and the curious nature of its hump have served to secure its place in the interest of natural historians from the time of Aristotle and Pliny the Elder. And the influence of the camel in human affairs cannot be denied – access to such singular physiology is almost wholly responsible for allowing the Arab nations to explore and trade within the heart of Africa, across the vast expanses of intimidating desert.

The camel inspires myth, but science has dismantled some of the common assumptions. The famous hump is no longer regarded as a water-storage facility but an accumulation of fat to be used as a source of energy in the face of dietary deprivation. The camel does not store water in any true sense at all; this idea probably arose from

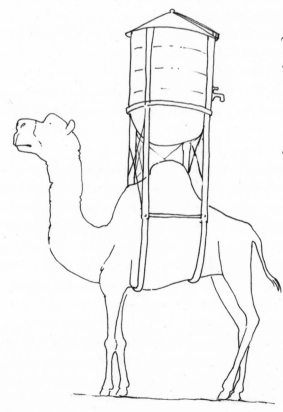

The camel does not store water in any true sense at all; this idea probably arose from its ability to drink up to 200 litres in less than 10 minutes.

its ability to drink at phenomenal rates – up to 200 litres in less than 10 minutes. Pliny thought that this huge volume of water was stored in the stomach for long periods, but modern observations show it to be almost completely absorbed into the system within 24 hours. This rate of absorption should, in theory, place the camel at serious risk of water intoxication. Rapid dilution of the blood ought to cause wholesale destruction of red blood cells from osmotic shock, as the cells absorb too much water and burst. But the camel, like the kangaroo, is equipped with unusually robust red blood cells that are resistant to deformation.

In the absence of water, the camel slowly but surely dehydrates and can lose 30 per cent of its body weight in this way without ill effects – compared to humans, where 10–15 per cent is usually fatal. Despite progressive dehydration, the blood volume remains remarkably constant and blood viscosity does not increase, thus avoiding the threat of heart failure. This is possible because fluid is withdrawn from the extracellular fluid spaces and not from cells or the vascular compartment.

The camel is able to adopt the same policy as the gemsbok in allowing body temperature to rise during the day, thereby limiting water loss. In a dehydrated state, body temperature may vary between a low of 34°C in the cool mornings to 42°C in the hot afternoons – the brain being protected by the carotid rete as in the gemsbok.

The difficulties of standing tall

There is really nothing quite like it in the world; it seems to belong more to the realm of the cartoon or of childhood fantasy than to reality. It is not difficult to imagine the astonishment and disbelief on the faces of the citizens of Rome when the first giraffe to be seen in a western cultural setting was paraded through the streets as part of the triumphal procession of Julius Caesar in 46 BC. The people were then required to wait another 300 years, to the time of the Emperor Gaudanius the third, before a further 10 giraffes formed part of the circus marking the first millennium of the founding of Rome. The perplexity of scholars at that time and their inability to come to terms with this unusual animal is reflected in its early name 'cameleopard' which attempts to marry its somewhat camel-like shape with its patchy markings.

Controlling blood pressure

The male giraffe stands 5.5 metres tall and weighs 800–1 900 kilograms. It possesses a thick, muscular tongue 45 centimetres in length, which appears immune to the pricks of the long, sharp acacia thorns the camel encounters as it browses on the green leaves that form its diet. Its great height provides an obvious advantage in allowing exclusive feeding rights to the crowns of trees, which are simply out of reach to smaller animals. It also permits a better visual appreciation of the surrounding plains, with the ability to detect the approach of predators from a considerable distance. But these benefits create engineering and physiological problems. How, for example, does the giraffe manage to deliver sufficient blood to its brain? And how is it able to breathe efficiently and fill its lungs when its windpipe (trachea) is two metres long?

To understand how the heart and blood vessels respond to these challenges, it is necessary to review, briefly, some aspects of blood pressure. When the heart of a mammal contracts it forces blood under pressure into circulation: the right ventricle pumps blood into the lungs where it is oxygenated, and the left ventricle pumps this oxygenated blood into the general arterial system to be distributed all over the body. As blood is forced into the arterial system, the pressure in the system rapidly rises to reach a peak (systolic pressure); and as the heart relaxes blood pressure falls to a lower level (diastolic pressure) before the next heart contraction increases pressure once again. Physicians express blood pressure readings as a ratio of systolic/diastolic pressure (measured in millimeters of mercury). In a healthy young adult, normal blood pressure is about 120/80. As age advances, the artery walls become less elastic and do not absorb or damp down the pressure wave as efficiently as in a younger person. As a consequence, blood pressure, particularly systolic pressure, increases with age. Blood pressure is controlled by a complex reflex arc, which involves specialised pressure receptors in the walls of certain arteries together with a nucleus in the brain stem and the nerves of the autonomic nervous system. This mechanism ensures that blood pressure remains within acceptable limits during such activities as exercise or change in body posture. Even after a substantial loss of blood, blood pressure can be maintained within the normal range for a reasonable time.

Arterial pressure drives oxygenated blood into organs and tissues to deliver essential nutrients. If delivery is impeded then the function of vital organs may be severely impaired. Organs vary in their sensitivity to low blood pressure but the brain is least tolerant in this regard – which poses a major dilemma for the giraffe.

The human heart pumps blood against gravity to the brain, but since the distance between the two is only a matter of some 35 centimetres, the amount of effort required is not great. As a consequence, the human left ventricle is rarely thicker than 11 millimetres, although it may be as much as 16 millimetres in a professional athlete. In contrast, the giraffe is faced with the task of pumping blood against both gravity and the hydrostatic pressure of a long column of blood for a distance of some 2.5–3 metres. To accomplish this, the heart must generate considerably more power – and this is achieved by increasing heart muscle mass. The thickness of the left ventricle averages 80 millimetres in the giraffe, almost eight times that of humans. To develop the necessary power to supply the brain, the giraffe's heart ejects blood under very high pressure. In the vessels close to the heart, blood pressure may be in the range of 280/150, but falls rapidly as the neck is ascended until, near the head, it is no more than 130/70. The blood vessels close to the heart are prevented from exploding under such high pressure by a coat of elastic tissue, which may be 12–15 millimetres thick (compared with two to three millimetres in humans), but which diminishes progressively as distance from the heart increases.

The circulation to the limbs faces the opposite challenge, because here the high pressure developed by the heart is now reinforced by gravity – so much so that blood pressure in the feet reaches a scary 350/180. To protect the arteries against explosion, the vessel walls are reinforced by a thick coating of muscle tissue so that they have the external appearance of thick cords with only a small central lumen for blood flow.

Bending down to drink should pose a major problem for the giraffe because the thin-walled vessels supplying the brain should, in theory, be exposed to the same high blood pressure found in the feet. If this really were the case then the animal would suffer a stroke every time it bent down to drink. Since this is obviously not the case it is likely that, during such a manoeuvre, there is constriction of blood vessels supplying the brain at a point before blood reaches the extremely thin-walled arteries traversing the carotid rete. This would have the effect of limiting blood flow and maintaining pressure at an acceptable level.

All of us have, at some time or another, experienced momentary dizziness when suddenly standing from a sitting or kneeling position. This sudden movement takes the reflex arc controlling blood pressure by surprise. As a consequence, blood pressure to the brain falls for a moment or two and causes dizziness. When the giraffe stands up after drinking, its head moves through a vertical distance of five metres in a matter of two to three seconds. So how much more of a problem should this create for a creature with a two-metre-long neck, when even humans, with hardly any neck at all, experience the effects of reduced blood supply to the brain. Such an action in the giraffe ought to be accompanied by an abrupt loss of consciousness. How the giraffe avoids this embarrassment is not well understood, but it may be explained, in part, by most of the blood destined for the head being diverted to the brain and away from other facial structures.

Cardiovascular control in the giraffe is extremely difficult to investigate experimentally, not least because of the size and bulk of the animal. It is also possible that structures inside the skull (the cavernous sinus and carotid rete) may play a major role in controlling postural change in blood pressure but, given their location, they are virtually impossible to access.

A question of breathing

If you can imagine breathing through a straw two metres long and three centimetres wide, which has been jammed into your mouth, you may have some idea of how it feels to be a giraffe. It is not so much whether you could breathe in such circumstances but rather whether you could breathe enough. The giraffe has the longest trachea in the animal kingdom and is faced with the problem of delivering adequate amounts of fresh air to the lungs down a long and relatively narrow passageway and expelling stale air back to the outside world. At any one time, the trachea contains about five litres of air and it was felt for many years that this huge 'dead space' must surely interfere with access to and elimination of air from the lungs where the real business of gas exchange occurs.

Research, however, shows that the giraffe deals with these apparent difficulties by the simple expediency of taking very large breaths. Watching a giraffe browse among the trees, you would hardly be aware of any breathing at all. Relaxed and at rest it breathes only eight to ten times per minute, but each breath takes in seven litres of air, some 14 times more than the average human adult. With each breath, air is sucked in at high speed down the trachea; at 180 centimetres per second, there is the danger of inducing wind-burn damage to the delicate lining of the respiratory tract. The high-suction pressure needed to drag in this amount of air might also be expected to result in inward collapse of the walls of the trachea. These difficulties are overcome firstly by mucus, which prevents wind injury, and also by a series of stiff, cartilaginous rings in the walls of the trachea, which provide structural support necessary to prevent tracheal collapse. These rings are present in all mammals but probably assume greater relative importance in the giraffe than in any other species.

'A giraffe is so much a lady that one refrains from thinking of her legs but remembers her as floating over the plains in long jambs, draperies of morning mist and mirage.' Such is the tribute of Karen Blixen to the grace and elegance of an animal that has excited and fascinated us for millennia – from its early depiction in the cave art of our ancestors to the more pragmatic considerations of the modern physiologist. From a scientific point of view the giraffe has given us genuine insight into the workings of the cardiovascular and respiratory systems, as well as thermoregulation. With its exotic form it has also shown us how evolution has adapted and fashioned these systems from the template of the basic mammalian model to create a unique and memorable animal.

Coping with adversity.

This chapter has presented only a smattering of the ways living things have adapted over the course of time to a whole variety of adverse circumstances. The various solutions which have been adopted are all, without exception, ingenious and it is difficult at first sight to avoid the conclusion that there has somehow been a degree of foresight and planning involved to secure a definite purpose. Natural though this interpretation may be, it is nevertheless incorrect. Natural selection is blind, it has no vision of the future but only the past hidden in the recesses of genetic material. If a chance mutation allows an organism to adapt better to the prevailing environment then it will flourish, as will its offspring. In this way, colonisation of new environments becomes possible where previously they were beyond the physiological and biochemical range of the ancestral population. But this is not a permanent arrangement for the simple reason that, as environments change, as inevitably they do, today's success stories become tomorrow's failures. At any one given moment in time we can marvel at how life copes with adversity but, in reality, these adaptations are engaged in a dance with time, where change is the rule and where evolution is ever destined to take life along novel and unexpected pathways.

If you can imagine breathing through a straw two metres long and three centimetres wide, you may have some idea of how it feels to be a giraffe.

CHAPTER TEN

ARE YOU RECEIVING ME?
reading the subtext

Human beings are compulsive communicators.

Compulsive communicators

If I want to talk to one of my children, scattered as they are around the globe, it is usually a simple matter of picking up the telephone and dialling the correct number. But other, more or less exotic channels are also available in the form of email, instant messaging via the Internet, Voice Over Internet Protocol (VOIP), facsimile, telegram, phone text or sms and the old fashioned but sadly underused medium of the personal, handwritten letter. Human beings are compulsive communicators with a seemingly insatiable curiosity concerning events in the world around them and the need to transmit this information to others, whether they live in the same house or thousands of kilometres away. Using words, we can express our thoughts, fears, wishes and intentions; and when language fails, when words cannot reach far enough to convey the true essence of our thinking, we can turn to the more abstract representations, which we know as art and music. In a more immediate sense, a simple gesture or a tone of voice may render a more precise meaning than words themselves. Whatever the format, the crux of communication is the dissemination of information, whether this be trivial, apocalyptic or something in between.

Although a dog has excellent vision and hearing, anyone who has ever taken one for a walk knows that it lives in a somewhat different world to our own. Gifted with a sense of smell several thousand times superior to ours, dogs possess a whole new way of perceiving their surroundings in ways largely denied to us, dependent as we are, for the most part, on our eyes and ears. Dogs need only sniff – the air, miscellaneous lampposts, shrubs, the bottoms of other dogs – for a quick summary of potential food sources, mating opportunities and the day-to-day traffic of its own kind in the neighbourhood. This sensitivity to what are essentially chemical messages is more important to the dog than any other sense, and this is reflected in the relatively large area in its brain devoted to perception and interpretation of olfactory stimuli.

The world is full of smells and, over time, evolution has fashioned systems enabling most living organisms to respond to them in one fashion or another. Of these, the most elegant is undoubtedly the use of chemical messengers in the construction of a system of communication.

What are pheromones?

Most animal species and many plants respond to chemical stimuli which we classify under the heading of odours. Aquatic life is no exception, but here the line between smell and taste is blurred and in many cases may amount to much the same thing. A chemical produced by one organism that results in some defined behavioural or physiological response in another member of the same species, *and no other*, is known as a 'pheromone'. Derived from two Greek words *pherein,* meaning 'to transfer' and *hormon*, meaning 'to excite', the word 'pheromone' came into usage to distinguish this specific form of chemical communication from responses evoked by a whole miasma of chemical signals in the environment. A hyaena can detect the smell of freshly spilled blood at a distance of five kilometres, a dung beetle can scent a fresh pile of elephant excrement from at least 300 metres and a salmon can locate the stream in which it was spawned by taste alone. But none of these examples make use of pheromones and there is no transmission of data from one to another member of a species.

Pheromones also need to be distinguished from hormones, which are chemicals, produced by an organism, that induce certain biological changes within that particular individual. In some cases, however, there

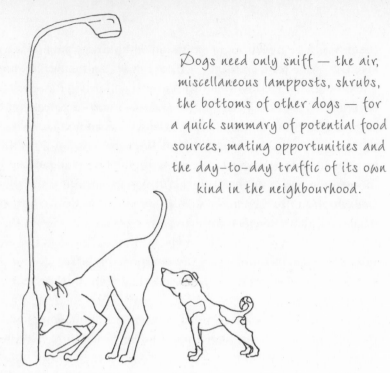

Dogs need only sniff — the air, miscellaneous lampposts, shrubs, the bottoms of other dogs — for a quick summary of potential food sources, mating opportunities and the day-to-day traffic of its own kind in the neighbourhood.

is a degree of overlap between the two. A lioness produces oestrogen, a hormone that prepares her reproductive tract for ovulation and pregnancy. She excretes some of this in her urine and a male lion, with the ability to assess by scent the concentration of oestrogen, can determine when she enters a state of full oestrus and is ready to mate. In this case oestrogen functions both as a hormone and as a pheromone.

Functions of pheromones

In single-celled organisms, pheromones represent the sole means of communication. In more complex animals a number of pheromones may be produced, each carrying a distinct message and supplying information as diverse as the threat of danger, sexual readiness, territorial boundaries, location of a food source or the wellbeing or integrity of a whole community. Such diversity requires that the message carried by each pheromone is specific and unequivocal – confusion, say, between an alarm signal and one of sexual attraction is likely to be 'embarrassing' at best and lethal at worst. Given the wide variety of circumstances under which they are employed, the physical properties of messenger molecules, such as their solubility and volatility, are critical to how they function as pheromones. A molecule, for example, that is insoluble in water would be useless as a pheromone in rivers, ponds and streams. Likewise, a leopard marking its territorial boundaries requires a pheromone that has low volatility and that will persist for several days or longer to provide a warning to possible intruders. These physical properties are influenced, in part, by the size (molecular weight) of the molecule itself – the lighter it is, the more volatile it is likely to be. The concentration of a particular pheromone may provide good information regarding the likely distance to its source – the higher the concentration, the closer the source – an important consideration in the case of alarm signals or sexual allurements.

Multiple pheromones in more complex societies

Complex societies such as those found among ants, termites and honeybees where there is a definite hierarchy and division of labour, require an equally complex system of communication. This is achieved through the agency of a whole raft of pheromones, some 30 of which have been identified in the honeybee alone. At the heart of honeybee society sits a single queen surrounded by large numbers of female workers who, at various times, may function as court attendants, foragers, builders, nursemaids, soldiers, porters, janitors, undertakers and general maintenance bees. Males (drones) are relegated to a lowly existence, their one useful function being to fertilise the queen during her nuptial flights. The queen is essential to the wellbeing of the whole colony and if she is removed, social disorganisation and collapse will quickly follow. From her mandibular glands she secretes 'queen pheromone', a complex cocktail of five different substances that is transmitted by touch to her courtiers and thence throughout the hive as the numerous bees come into physical contact with one another. Queen pheromone serves to inform the colony that she is alive and well and in residence. As it is carried to the further reaches of the hive, the concentration of queen pheromone diminishes and, in a large colony, it may be essentially non-existent by the time it reaches the edges. This may act as the stimulus for the colony to divide into two or more groups and thus provides a mechanism by which the size of a given colony may be limited.

The queen does not entertain competition during her regal tenure and to this end she secretes a 'footprint' pheromone from her tarsal glands and a tergite gland pheromone from her tergal gland, which respectively inhibit the rearing of a second queen and suppress the development of functional ovaries in the rest of the workers. Only when the queen dies will ovarian maturation occur in other workers and pave the way for the construction of queen cells within the hive. Special nutrients fed to the larvae in such cells will see the emergence of a number of potential queens. Open competition between these, with fights to the death, eventually results in the succession of a single queen.

Other pheromones have been shown to operate in activities such as food gathering, defence, nurturing of larvae and the growth of the colony, but detailed mechanisms are incompletely understood. However, two pheromones that are reasonably well understood are those concerned with aggressive defence behaviour and removal of the dead. During early decomposition, a dead bee produces oleic acid (a long chain fatty acid), which functions as a posthumous pheromone stimulating other bees to remove the corpse from the hive. So embedded is this behaviour that a live bee coated artificially with oleic acid (in the interests of research) will be promptly evicted from the colony despite demonstrating with clear and vigorous physical protestations that it is, in fact, very much alive.

When danger threatens the community, workers guarding the entrance to the hive start to secrete isopentyl acetate, a highly volatile pheromone that is rapidly dispersed through the colony by the guards' vigorously beating their wings. This alarm signal is produced by glands adjacent to the sting chamber – a fact that can have unfortunate consequences for victims of bee stings. When the barbed sting penetrates something as tough as human skin, the barb cannot be withdrawn and, as the bee flies away (to die), the whole sting assembly and associated glands are left behind. The alarm pheromone continues to be generated by these glands for a little while and attracts other bees to sting in the same location – a matter of some concern if you are unlucky enough to irritate a swarm of bees.

Sexual attraction

Outside the emotional sphere of human sexual attraction, sex in most simpler forms of life is governed by biochemical imperatives. The great need of nature and natural selection to bring together male and female gametes in sexual reproduction is such that it cannot be left to the whim of the individual. Asexual reproduction may provide a fail-safe mechanism, but it results in a static and unchanging population, ill equipped to face unexpected shifts in environmental conditions. Chemical attraction between sexes, mediated through pheromones, promotes sexual reproduction and hence the possibility of genetic variation.

As we have discussed in Chapter 7, sexual selection is frequently determined by a hierarchy of dominance where open competition between males results in the emergence of a single male with exclusive access to breeding females. Effective though this may be, it is nevertheless the female who usually commits to a greater reproductive investment, dedicated as she is to feeding and protecting her young. In accepting a mate she may therefore use other more subtle indicators in assessing the desirability of a particular individual. This may come, for instance, in the form of recognising that a particular male controls a territory with higher quality resources than the territories of competing males. The red-backed salamander marks his territorial boundaries with faeces and pheromones – the odour of which reflect the quality of his diet. From this odour the female obtains a measure of the quality of the habitat controlled by this particular male and, if it is highly satisfactory, she is more likely to mate with this particular animal, thus acquiring good quality food resources for both herself and her offspring.

In some cases the female may assess the suitability of a male by the quality of the gifts he brings to her. The males of some species of insect such as the queen butterfly found in Florida and the North American tigermoth are known to feed on plants containing high levels of alkaloids – powerful vegetable poisons secreted by plants to protect them from the attentions of leaf eaters. But these insects, along with some related species of butterfly, appear to be immune to the toxic effects of eating such plants and use the toxins as a defence strategy against predation. Predators have learned to avoid these insects because of their bitter taste and poisonous character, so the presence of alkaloids confers a definite benefit in terms of survival. The females of these species prefer to mate with males containing higher alkaloid levels, since these compounds are transferred to her during copulation and not only increase her own chance of survival but also that of her eggs and larvae. Pheromones may function as long-distance sexual attractants, providing a means by which widely dispersed members of a species can come together. In the 19th century French naturalist Jean-Henri Fabre noticed that a female emperor moth, which he kept in his study, attracted large numbers of males even when her enclosure was covered by a cloth but not when this was rendered airtight. This observation suggested the release of a volatile chemical attractant by the female, although initially some scientists believed the emission of infrared radiation to be responsible rather than a chemical message. After 20 years of painstaking research, Adolf Butanandt – a German chemist and Nobel Prize winner – finally identified the attractant pheromone of the female silkworm moth. From over half a million moths he was able to extract a mere 6.4 milligrams of pure pheromone, which he called *bombykol*. The male moth is able to detect this substance in incredibly low concentrations (1×10^{-18} grams) at distances of over two kilometres using microscopic hairs on its antennae.

In the absence of the direct introduction of eggs and sperm by copulation, reproduction in water produces many difficulties. The presence of water currents combined with rapid dilution greatly reduces the chances of external fertilisation. Yet this is the method adopted by many fish, which have solved the problem to some extent by ensuring the perfectly synchronised release of eggs and sperm. In the evening hours, the female goldfish produces a steroid-like hormone known as 17,20P, which is responsible for the final maturation of her eggs. As the concentration of this substance increases in her blood, some is excreted into the surrounding water. The male goldfish is able to detect this by smell and it causes him, through an indirect action on the brain and pituitary gland, to increase the production of sperm. The following morning when her eggs are fully developed, the female releases two pheromones belonging to a group of chemicals known as prostaglandins which act as sexual allurements to the male. In response the male chases and nudges the female until she eventually releases the eggs, which he immediately covers in sperm, thereby guaranteeing a high rate of fertilisation. In this instance the steroid 17,20P acts both as a hormone and a pheromone.

Alarm pheromones

There is a clear advantage to the survival of a species if one of its members under threat can alert others of its kind to either flee or mount an appropriate defence, even though this may do little to prevent damage or death to the individual being attacked. The sometimes heavy price exacted for this altruism dictates that alarm signals are directed primarily towards close relatives or to clones (where asexual reproduction produces identical genetic replicas) rather than distant relations or even members of another species.

The clonal sea anemone spends its life attached to rocks in the waters off the Californian coast. Colonies form tightly packed groups and reproduction is asexual, with individuals dividing longitudinally to produce exact replicas of themselves. Food drifting on currents is captured by entanglement in their tentacles and is directed to the mouth. Unfortunately for the anemone, its tentacles are the favourite food of a certain marine slug. In response to assault by this predator, the anemone withdraws its tentacles while at the same time releasing a pheromone – anthopleurine – that diffuses through the surrounding water to alert other anemones, which in turn respond by bodily contraction and withdrawal of tentacles. Given the rapid dilution of this pheromone and the vagaries of water currents, this mechanism serves to alert anemones only in the immediate neighbourhood. However, the sea slug itself becomes an unwitting early warning system as the anthopleurine it has ingested eating sea anemone tentacles is excreted from its body, thus giving advanced notice of its arrival wherever it may travel.

Karl von Frisch, who received the Nobel Prize for his description and elucidation of the dance of the honeybee, was also instrumental in drawing attention to the probable use of chemical messengers in initiating fright responses in fish. In the course of some experiments with European minnows, it became necessary to identify individual fish, which he achieved by making a small incision in the skin. Returning such a wounded minnow to the aquarium caused other fish to flee from it in apparent alarm and von Frisch postulated the release of an 'alarm substance' that spread through the water alerting others to danger and provoking escape behaviour. This chemical trigger has now been identified as a derivative of hypoxanthine, which is produced by club cells in the skin of minnows and other related species. Under normal conditions, the pheromone produced by the club cells has no access to the external world and it is only when the skin is mechanically damaged, as it would be by a predator, that the pheromone is released. Carried on the current, the pheromone travels downstream, where it stimulates other minnows to take evasive action. All is not lost for the fish that happen to be upstream of the event, since they are able to respond to visual cues and react accordingly when they witness the sudden departure of their fellows.

Pheromones that mark trails and boundaries

Alarm pheromones, such as those employed by the sea anemone and the minnow, disperse rapidly when the period of danger has passed – failure to do so would mean that, over time, the pheromone would either cease to have any relevance as a warning system or else would create an entirely neurotic population believing itself to be under constant threat. But in the case of trail and boundary pheromones, persistence of the chemical messenger is of paramount importance, particularly where the demarcation of a territory is concerned. Given that small molecules disperse more rapidly in air, pheromones used in marking trails or boundaries are represented mainly by heavier molecules. Molecular size also varies with the background environment: under the conditions of higher temperatures and humidity encountered in tropical forests, it makes sense to use a pheromone with a heavier, more stable structure compared to, for example, a pheromone carrying out essentially the same purpose in the cooler, drier regions of a temperate grassland where dispersal through evaporation is less of a problem.

Laying a scent trail identifying the location of a food source represents a major form of communication to social insects such as ants and termites. Ants secrete a trail pheromone from a small gland (Dufour's gland) located at the base of the sting. Returning to the nest after discovering a source of food, a worker ant periodically touches its rear end to the ground to deposit minute amounts of pheromone, which thus serve as a trail for others to follow. The quality of the food source is identified by the frequency with which the ant stops and marks the trail. Presented with either a strong or weak solution of sugar, ants were found experimentally to lay almost 50 per cent more trail marks from the stronger solution. So sensitive are ants to trail pheromones that estimates show the total pheromone output from a single nest of the leafcutter ant (one milligram) would be sufficient to lead a column of ants three times round the circumference of the Earth. By nature, ants are secretive creatures so it is not surprising to find that trail pheromones are specific to each species.

The use of pheromones to establish the limits of a territorial boundary is widespread throughout the animal kingdom, particularly in mammals and other land-dwelling vertebrates. Usually deposited in urine or faeces, but sometimes from glands adjacent to the eyes, these pheromones function as 'keep out' signs in the physical absence of the owner of the

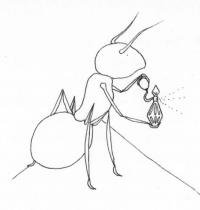

Laying a scent trail identifying the location of a food source represents a major form of communication to social insects such as ants and termites.

territory and are usually directed at males of the same species. A complex mixture of pheromones and excretory products gives to any potential intruder information regarding the degree of dominance and state of health of the territory's owner and hence the likely consequences of trespass. Markers containing pheromones from a number of individuals are found at common boundaries or when the ownership of a particular area is disputed. The most dominant animals occupy the best territories – as defined by quality of food supply and access to breeding.

Although marking a territory theoretically confers absolute right of ownership, this is seldom realised in practice since the owner cannot be everywhere at once. This probably explains the inordinate amount of time spent by male lions in patrolling and marking boundaries. But even here, lion behaviour can be very flexible, and research scientist Pieter Kat in his study of the Okavango Delta lions has, on several occasions, seen the presence of nomadic males being tolerated by the pride males, even to the extent of feeding on the same carcass. Since on most occasions the nomads are vigorously repelled, this sporadic tolerance remains unexplained. Perhaps it is only permissible when the females are not in an oestrus state or perhaps violations may be allowable on the peripheries of territories but not in the core areas where prey density is likely to be highest.

Scent marking is not unique to territorial mammals and is found also in the South American rodent, the mara, where the male sprays urine onto the female. This behaviour is also seen in goats which rub the products of their urine onto the backs of females when mounting. Presumably this behaviour is designed to discourage other males and may represent a form of chemical guarding when the female is approaching oestrus.

Pheromones as agents of deception

Plants have devised a number of strategies to ensure successful pollination, the colour and odour of flowers being designed to attract insects that unwittingly perform as vectors in the transfer of pollen from one plant to another. Not all plant odours are pleasant and some produce perfumes strongly resembling those of faeces or putrefying corpses in order to attract their preferred pollinators – dung beetles or insects that normally feed on carrion. Rather than rely on colour or smell, some plants, particularly the orchid family, have flowers that, in appearance, mimic almost precisely the females of certain insect species. The male is deceived into landing on this pseudo-female with a view to copulation but instead picks up a pollen package, which he duly transfers to the next flower. Some orchids, such as the genus Ophrys, which grows in Mediterranean regions, have taken a further quantum leap in this art of deception. In a feat of evolutionary wizardry, they have not only succeeded in imitating the external appearance of the insect pollinator but have also achieved the synthesis of molecules very close in structure to those of the natural sex pheromones of the pollinators. Specialisation may be so specific in some, that only a single species of insect is attracted to the flower, while in others a pheromone cocktail seems to be produced that has a wider appeal to a variety of bees and wasps.

On a more aggressive level, a group of spiders known as bolus spiders have evolved the ability to synthesize the female sex pheromones of certain moths on which they prey. The male moths have little choice but to home in on the sexual signal and may then be captured in a rather unusual way. Bolus spiders do not spin webs in the ordinary sense but instead weave a sticky silken ball, which is suspended on a single silken thread. As the moth comes within range the spider swings this silky ball like a clock pendulum and directs it at the moth, to which it sticks firmly if the aim is accurate. Surprisingly enough, this method of securing prey is as successful as the more usual orbital web found in other species.

Two ways of smelling

Amphibians, snakes, lizards and most four-footed animals, but not humans and other primates, possess a remarkable duplication of structures designed to interpret the world of smells. The one most familiar to us consists of scent-sensitive receptors located in the lining of the nasal cavities. Nerve fibres from these receptors pass to an area at the front of the brain, the olfactory area, where various smells become translated into consciousness. But there is also a second system located in a structure called the vomeronasal organ (VNO) which is a blind-ending sac connected by a small channel to either the roof of the mouth (snakes and lizards) or to the nasal cavity (most mammals). The nerve fibres pass from here not into parts of the brain associated with conscious perception but relay instead to the limbic system – an old part of the brain in evolutionary terms that is concerned primarily with emotional behaviour and motivational drives. The hypothalamus is part of the limbic system and this structure, as we have seen, is intimately linked with the workings of the whole endocrine system. Thus it is possible that pheromonal odours processed by the VNO can initiate alterations in endocrine function, most particularly in matters concerning reproduction.

The primary olfactory system can be seen as the mechanism that leads to the conscious detection of odours, while VNO-mediated olfaction, connected as it is to the endocrine system, can be regarded as an unconscious pathway suited better to processing information specifically related to pheromones rather than general environmental odours. Although a plausible explanation, this unfortunately does not appear to be wholly correct. Female sheep, for example, will still produce a surge in reproductive hormone output in the presence of a male even though all the nerve connections to the VNO have been surgically sectioned. In this case the odour of the male and the response to it seem to have been mediated through the primary olfactory system. Notwithstanding this and other examples, it may well be that the VNO does function as the most important pheromone receptor, but simply does not enjoy exclusive rights. Some biologists believe there is some type of interaction and interdependence between these two modalities of olfaction, though the precise details of how this may work appear obscure at the present time.

The ability to transfer pheromones from the outside world to the VNO has led to some interesting adaptive behaviour. In mammals, where the VNO is connected to the nasal cavity, it may be a simple matter of breathing through the nose; but some species, notably the cat family, ungulates (hoofed animals) and marsupials, engage in a ritual that has been termed 'flehmen'. During investigation of the reproductive status of a female, males of these groups sniff and then lick her urine. This is followed by drawing back the lips into something representing a grimace, tilting the head backwards and breathing deeply. The purpose of this rather extraordinary behaviour is to transfer odours directly to the VNO where pheromone signals can trigger appropriate mating responses.

By flicking the tongue in and out, snakes and lizards are able to transfer airborne pheromones directly to the opening of the VNO located in the roof of the mouth. Secreted in the form of scent trails, pheromones produced by snakes may be essential in the business of finding food and courting a mate.

Pheromones in a mammalian setting

Mammalian pheromones are largely confined within the spheres of reproductive behaviour, demarcation of territorial limits and recognition of family members. They are produced by a variety of glandular structures and are discharged to the outside world by means of urine, faeces and skin secretions.

Involving as it does several hundred thousand animals, the annual migration of wildebeest over the Serengeti Plains in northern Tanzania represents the largest movement of a single species of mammal on Earth. Somewhere in the midst of this seething horde are several thousand youngsters, dependent on their mother's milk for survival. Unlike lion cubs, wildebeest calves are denied access to any lactating female other than their own mother, so recognition between mother and offspring is critical and, one imagines, potentially the stuff of ungulate nightmares. This problem has received little scientific scrutiny in the wildebeest but, presumably, it is similar to that faced by sheep, another animal that gathers in large numbers and suckles only its own lambs.

Immediately after giving birth, a ewe will sniff and lick her lamb, including the amniotic fluid that may still cover it. This innate behaviour, together with the vaginal stimulation associated with delivery, sets in motion a complex chain of neurochemical events in the mother, which ultimately leads to the secretion of the hormone oxytocin from her pituitary gland. Oxytocin acts on the olfactory area in the brain in such a way that the odour of the lamb becomes firmly imprinted on her. Remarkably, the sheep is able to learn and fix this scent in memory in a very short space of time, usually no more than two to four hours, though the precise nature of the substances responsible for this odour signature is as yet unknown.

By hook or by crook, it is the destiny of every male to try as hard as he may to transmit his genes onward to the next generation. To this end he invests a good deal of time in ascertaining the readiness of females to mate. Oestrus, as the prelude to ovulation, is accompanied by an increase in the production of the sex hormone oestrogen, some of which is excreted in urine and which the male can detect and quantify via vomeronasal olfaction. In this context oestrogen functions as a pheromone, advertising both the receptivity and fertility of the female and thereby increasing the likelihood of a successful reproductive outcome. This method is a widespread and relatively simple means of communication employed as a reproductive strategy in mammals. In some species, however – for example mice and hamsters – a number of discrete steps can be recognised as precursors to mating.

The golden hamster is a species that seems to be highly dependent on pheromones for successful reproduction. In a series of experiments at Rockefeller University in the USA, William Agosta and his team were able to identify and isolate a number of substances with pheromone properties from the vaginal discharge of the female hamster. The first of these, dimethyl sulphide, functioned as the basic attractant pheromone drawing the male to the female; and so sensitive was this message that the researchers estimated only 200 molecules were necessary to cause visible excitement in the male. Having found the female, the male hamster proceeds to investigate her further by smelling and licking her ears, head, flank and genital region. Only after completing this latter manoeuvre does he attempt to mount her. Strangely, this mounting behaviour was able to be duplicated by painting the vaginal discharge onto the rear end of another, anaesthetised, male, and

The male hamster proceeds to investigate her further by smelling and licking her ears.

raised the possibility that this material contained a distinct 'mounting pheromone'. After much further work, researchers were able to isolate a protein which they termed aphrodisin, capable on its own of initiating mounting behaviour. This was the first time that a molecule as heavy as a protein was recognised as a pheromone. These experiments also confirmed the importance of the VNO, since the whole behavioural sequence was abolished if the VNO was rendered non-functional.

The production of mammalian sex hormones is, however, not the sole prerogative of the female. In domestic pigs, the saliva of the boar contains a steroid, 5 alpha-androsterone, which causes an oestrus sow to adopt the stance normally seen during copulation. As it happens, this chemical agent is also produced by truffles, a highly prized gourmet food item, which explains why pigs are so much more successful than other animals in finding them beneath the soil of the oak woods of France.

There is a remarkable interplay of sex pheromones in the social organisation of mice, mammals well known for their large communities and prolific breeding. In a society where all the females breed (but only some of the males), checks and balances are required to prevent a colony from outstripping its available food supply. To this end a variety of pheromones are released in the urine of both males and females which alter the reproductive dynamics of the community. Thus it is that, when food is plentiful, the mere presence of an adult male will cause mature females to go into oestrus and immature females to undergo accelerated puberty. Observations also show that the presence of a strange male not only prevents the implantation of an ovum recently fertilised by another male, but also results in the rapid return to oestrus. From the point of view of the stranger, this effect provides him with the opportunity to pass on his own genes at the expense of local competition. At the same time there is benefit to the society as a whole, the genetic variation ensuring a more robust population.

Under the artificial conditions of the laboratory, females crowded together in the absence of males are able to suppress oestrus and delay the onset of puberty in younger animals. The pheromone responsible for this has been identified as dimethylpyrazine, though its role under more natural conditions is uncertain. However, in the experimental conditions of female overcrowding, some advantage accrues in that females do not have to waste metabolic energy in what is essentially pointless oestrus. Faced with diminishing resources in the natural world, the females of some species secrete a pheromone that alters the sex ratio of offspring. In normal circumstances the distribution of the sexes is, to all intents and purposes, equal; but, as food supplies run out, the number of males may increase to about 70 percent. Presumably this works on the premise that if reproductive behaviour cannot be wholly suppressed, then the next best thing is to produce fewer females, resulting in fewer offspring and hence less competition for food.

Pheromones in humans

Our perception of the outside world comes mainly through sight and hearing. Humans and the great apes do not possess a vomeronasal organ, which makes only a brief appearance during embryonic development as an evolutionary relic. Our sense of smell is the least developed of the five senses and, though desirable, it is not essential to the life of the individual. Nevertheless, scent has featured as an important part of most cultures since the very beginnings of civilisation. The experience and interpretation of a whole range of smells depends, in large measure, on the cultural setting, which makes it difficult to separate out the biological importance of a given odour from its social context. Over the past few years, in particular, western society has been seduced by advertisers into the belief that a certain scent confers on its wearer an irresistible attraction, whether this is in a romantic or other setting. Against this background, it is reasonable to ask what relevance human pheromones might have to a species that has developed such complex social and cultural interactions. But this, of course, is to assume that humans produce pheromones in the first place.

Menstrual synchrony

The first glimmer of proof regarding the existence of human phero-mones came from a study conducted on the menstrual cycles of under-graduates living together in a college dormitory. Among the population of 135 students, American psychologist Martha McClintock discovered that, during the study period October to April, synchronisation of menstruation occurred among close friends and room mates but not throughout the dormitory as a whole. In trying to account for this phenomenon she was able to exclude a number of possible factors such as similar diets, lifestyles, awareness of each other's cycles and even the length of exposure to light and dark. The single common factor

137

appeared to be the length of time the young women spent together: the longer the time, the greater the degree of menstrual synchrony. Was it possible for a pheromone to be responsible for these events?

When this work was published in 1971 it was not greeted with universal applause and seemed to have been almost forgotten until 1986 when organic chemist George Preti and his colleagues in the USA not only confirmed McLintock's original observations but also provided evidence that a pheromone was indeed involved and that its likely source was the axillary sweat glands.

Sweat glands and body odour

Humans possess two distinct types of sweat glands. Eccrine sweat glands are distributed all over the body and secrete a clear, odourless fluid containing variable amounts of salt. This is the fluid we generally know as sweat and, under tropical conditions, we may produce up to 3 litres an hour. Evaporation of this sweat on the skin surface causes cooling, which maintains body temperature within closely defined limits – a process known as thermoregulation.

Apocrine sweat glands develop fully only during and after puberty. They form part of the secondary sexual characteristics of both men and woman and are located only in the axilla, genital region and nipples. Only a very small quantity of secretion is produced but this is largely responsible for the generation of body odour. Secretion is increased at times of sexual arousal and emotional stress. Both eccrine and apocrine secretions undergo bacterial decomposition in the warm conditions found in the axilla and genital area and this undoubtedly contributes significantly to the odour profile.

Following on the implication of pheromone involvement in the menstrual history of women, McClintock, in collaboration with Kathleen Stern, extended her work and was able to demonstrate convincingly the likelihood of two *distinct* pheromones being released during different stages of a normal cycle. In addition, exposure to the axillary sweat of other women was shown to advance or delay ovulation. These experiments illustrate scientific methods so well, they are worth considering in more detail. By way of introduction we need to know that each menstrual cycle has three parts: (a) menstruation; (b) a follicular stage following menstruation, characterised by rebuilding of the endometrial lining of the uterus and maturation of the ovum in the ovary to the point of its release at ovulation; and (c) a luteal phase following ovulation where the now empty ovarium follicle forms a structure called the corpus luteum, which secretes the hormone progesterone and prepares the uterine lining for implantation of the fertilised ovum.

Twenty young woman aged between 20 and 35 years volunteered for these experiments, all of whom were healthy and had regular menstrual cycles. Their urine was analysed on a daily basis to detect the peak levels of luteinising hormone, which coincides with ovulation. Nine other woman with similar characteristics served as donors of sweat. Each of these nine was asked to wear a cotton pad under one armpit on a daily basis for at least eight hours, having previously avoided all perfume products and certain foods. The pads were then removed, treated with Isopropyl alcohol (a strong-smelling substance that masks any possible conscious appreciation of sweat-derived odours) and frozen in glass vials. The precise phase of the menstrual cycle of the donors was known at the time of collection of the pads (by urine analysis). After thawing, the pads were wiped across the upper lip of the volunteers, who were advised not to wash their faces for six hours. Pads from donors in their follicular phase were applied daily to the volunteers for a period of four months followed by a further four-month exposure to pads from the luteal phase.

The net result of these experiments showed that women exposed to the sweat of someone in her follicular phase developed shorter menstrual cycles, while the opposite occurred with exposure to sweat from the luteal phase.

What we can glean from this experimental data is, firstly, that there is clearly something in the sweat of women that is capable of altering the menstrual cycle length of others. The most likely candidate for this is a pheromone, given that it is able to regulate a neuroendocrine mechanism in someone who cannot consciously perceive it as an odour. Secondly, if we accept this pheromonal basis, there must be two of them to account for the opposite responses to sweat taken at two distinct phases of the cycle. Thirdly, it is clear that the timing of ovulation can be chemically manipulated.

If the sweat of one woman can influence the menstrual cycle of another, what effect might men have on women? With an experimental design very similar to the one outlined above, it was shown that women who had previously experienced marked irregularity in their menstrual cycles suddenly developed cycles of almost clockwork regularity when exposed to male sweat. And more recent research suggests that men may unconsciously be aware of the time of ovulation in women with whom they have close contact.

Fascinating though these observations may be, it is reasonable to ask at this point what relevance these pheromones have to the lives of modern men and women. After all, humans are not seasonal breeders, nor do they limit sexual activity to oestrus phases as seen in other mammals. The short answer is that they probably have very little functional significance today, but conceivably represent an important part of our ancestral past, when our progenitors were living as troops of primates in the forest.

Pheromones conceivably represent an important part of our ancestral past, when our progenitors were living as troops of primates in the forest.

140

Rather than synchronisation of menstruation, it may be the synchronisation of births that is the crucial event. In an ancestral population continuously on the move, a co-ordinated birthing of offspring offers both the possibility of sharing maternal duties and the opportunity for youngsters to learn appropriate life skills together and thereby develop stronger social bonds. It may also serve to keep the troop together, rather than leaving behind stragglers with young infants. But this is simply speculation and, in truth, the rationale behind menstrual synchrony is completely unknown.

Sweaty T-shirts and sexual attraction

The immune system of humans – and of all mammals – recognises when the body is invaded by foreign material such as bacteria, viruses, parasites or a transplanted organ. In response, it deploys white blood cells to the site of the intrusion to seek out, attack and destroy the alien presence. To perform this task it must be able to distinguish between 'self and non-self', a function residing in a group of some 82 genes localised on one of the chromosomes contained in white blood cells and known collectively as the 'major histocompatibility complex' (MHC) or human leucocyte antigen (HLA). This ability to tell the difference between 'mine and yours' is possible because each organism carries its own unique biological blueprint within its genes. The success of a kidney transplant depends on ensuring as close a match as possible between the MHC types of donor and recipient – recognising that the only perfect match occurs between identical twins.

The odour associated with the urine of mice has been known for many years to vary according to MHC type, although the mechanism by which the genetic makeup is translated to odour is not fully understood. Nevertheless the differences in odour are so marked that variations in only one gene sequence in MHC can be detected. But what possible biological use might this represent? Laboratory studies demonstrate that mice prefer to breed with individuals whose MHC type differs from their own – a factor they can identify by smell. Consequently, they avoid mating, as far as possible, with their own siblings and cousins. Inbreeding brings with it the risk of genetic uniformity and hence the increased likelihood of failure to adapt to changing environmental circumstances. It also increases the risk of adverse genetic mutation, so rejecting a mate with a similar MHC type should lead to a healthier population with a wide genetic diversity.

If mate selection in mice can be influenced by MHC type, as expressed through body odour, might the same apply, in some measure, to human mate selection? This intriguing possibility was examined by biological researcher Claus Wedekind and his colleagues in Switzerland. Under carefully controlled conditions, a group of young female students was asked to grade, in terms of intensity and pleasantness, the odour of plain cotton T-shirts worn overnight by a group of male students. Although women preferred the odour of MHC dissimilar men, this only applied if they were not taking oral contraceptives; women on the pill expressed preference for MHC similar men. There is no definitive explanation for these observed differences. Oral contraceptives mimic, to a large extent, the hormonal profile of pregnancy and an explanation of sorts suggests that women in this hormonal state may have a different agenda, being less interested in mate selection and more responsive to the companionship of closer kin – as expressed by a similar MHC type.

Further T-shirt experiments showed men to prefer the odour of MHC dissimilar women. More recent work from the same research group indicates perfume selection by women is partly determined by MHC type, suggesting the possibility that perfume may function as an amplifier or enhancer of MHC-related odours. And finally, research among married people in an isolated community in western Canada revealed a higher proportion of marriages between MHC dissimilar types than would be expected by chance alone, though these findings have not been confirmed elsewhere.

The authors of this research have been at pains to point out that the selection of a suitable partner, perhaps for life, cannot be attributed simply to a matter of smell. What these experiments do unequivocally establish is a definite relationship between genetic makeup and body odour and the ability of both sexes to discriminate between individuals on this basis. Whether they act on this information is a different matter altogether. Mate selection is so overlaid by social and cultural factors that olfactory ones may be pushed into the background; but just how far they are pushed remains to be seen. Despite the relative poverty of our olfactory sense compared to that of other mammals, our sense of smell still functions as an important asset. Smell forms the basis of an odour-associated memory, which may be more emotionally evocative than any other form of memory, and it enables mothers and infants to form possibly the closest bond in Nature.

CHAPTER ELEVEN

WHAT IS LANGUAGE?

communication in the bush

*Speech it may
not be, but
communication
it most certainly is.*

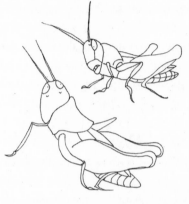

Vocal communication

To suggest that 'language' is used as a means of communication among animals other than humans is, in some circles, to commit scientific heresy and invite derision. If we believe the use of words to be the defining criterion of language, then it is clear we have essentially closed the door on all species other than our own. The structure of the human larynx is unique in the animal world and enables us, with help from the tongue, lips and pharynx, to articulate thoughts and ideas in a way that is unavailable to any other creature. But this does not exclude the possibility of other animals making use of sound as a vehicle of expression: speech it may not be, but communication it most certainly is. At a simple level, the clicks and chirpings of grasshoppers, crickets and cicadas may function as indicators of aggression, sexual attraction or a call to encourage aggregation. It is, however, only in some species of mammal that the repertoire of sounds is sufficiently broad and complex to be considered as representing a form of language. An excellent example is provided by the vocal range of elephants in relation to various aspects of their behaviour.

Infrasound communication

Katherine Payne, a biologist at Cornell University in New York, had spent some years researching communication in baleen whales and had found that calls between whales were largely within the range known as infrasound. Low-frequency sound, below the threshold of human hearing, is part of the background noise of life, fuelled significantly by events such as earthquakes, volcanic eruptions, thunderstorms and the impact of waves on the shore. Standing one day at the elephant enclosure at the Washington Park Zoo in Oregon, Payne became aware of a palpable throbbing or vibration in the air, which she likened to the sensation of standing in the vicinity of the largest organ pipe during a church service. Being familiar with the phenomenon of infrasound transmission, it occurred to her that perhaps the elephants were employing this as a means of communication. During October 1984, she and her colleagues made 45 hours of sound recordings at the zoo, followed by a further 15 hours at Circus World in Florida. Analysis of the data confirmed that as many as one-third of the calls made by elephants were below the threshold of human hearing. Hundreds of calls had fundamental frequencies in the range of 20 cycles per second or lower (usually expressed as Hertz or simply Hz) – the human ear detects frequencies between 25 and 20 000 Hz, though in the lower range sound is perceived as vibration rather than noise.

Elephant talk

Botswana is home to a large number of elephants, and a quiet afternoon spent at a waterhole by the Chobe River may often be enjoyed in the company of 200 or more of them. In addition to providing liquid refreshment and the ecstasy of a mud bath, these occasions are important for social interaction. Family groups, large and small, emerge from the bush to engage frequently and enthusiastically with others in a loud chorus of rumbles; some intertwine trunks, youngsters chase each other in and out of the water and adolescents push and shove in the manner of adolescents everywhere. The air may resonate with trumpet blasts, screams, rumbles, snorts and bellows. To even the most casual observer, it is obvious that these are not simply random noises.

Biologists who have studied elephants over many years have often been perplexed by some aspects of their behaviour. A family group moving across an open plain may suddenly freeze in its tracks, only to set off in an entirely new direction a few minutes later. Rowan Martin, a researcher in Zimbabwe, placed radio collars around the necks of selected dominant adult females, and discovered that the movements of different families were frequently co-ordinated, often for several weeks at a time, even though the groups were several kilometres apart with no possibility of visual or olfactory contact.

Following up on her discovery that infrasonic communication is used among captive elephants, Katherine Payne set about broadening her field of research, and travelled with her recording equipment to join the Amboseli Elephant Research Project in Kenya, to work alongside Cynthia Moss and Joyce Poole. One thousand hours of recording out on the plains confirmed Payne's original observations: that most of the calls emitted by African elephants fell within the same 14–35 Hz frequency range as their Asian counterparts in the Washington Zoo; and once again, only one-third were audible to human observers, even to the ears of skilled elephant researchers such as Joyce Poole.

The elephant repertoire

From these recordings, together with direct observation of elephant behaviour associated with their calls, it was possible to identify a repertoire of 30 distinct calls – both below and within human range – including:

Sexual excitement	Social excitement	Distress
oestrus rumble	greeting rumble	lost call
female chorus	social rumble	suckle rumble
genital testing	roar	suckle cry
	mating	distress call
	pandemonium	reassurance rumble
	play trumpet	calf response
	social trumpet	suckle distress scream

Group dynamics	Fear, surprise or strangeness	Dominance
attack rumble	trumpet blast	female – female
let's go rumble	snort	male – male
contact call	scream	musth rumble
contact answer	bellow	
coalition rumble	groan	
discussion rumble		

(From Poole 1996)

When elephants feed, which is most of the time, a family group may be spread over a relatively large area of woodland. Reassembling as a group always elicits a series of greeting calls and, the longer the separation, the longer will be the duration of calling. The matriarch may then emit a 'let's go' rumble, calling the group to move to a different location. If some of its members are reluctant to move, a good deal of communal rumbling may ensue, passing back and forth between individuals as a 'discussion' rumble, until some form of consensus has been reached. In general, females call much more frequently than males, but this changes when a bull enters musth and actively seeks out an oestrus female. Since oestrus occurs only once every four years or so (two years of pregnancy plus two years of nursing) and lasts three to four days, it is important she find a suitable mate without delay. A bull elephant crisscrossing a territory in search of a female will find his 'musth' rumbles answered by 'oestrus' calls from the female. Copulation, when it occurs, is followed by great excitement in elephant circles, the mated female producing a series of low-frequency post-copulatory calls for up to 30 minutes, while other family members engage in rather bizarre 'mating pandemonium' involving rumbles, screams, trumpeting, temporal gland secretion and defecation.

Because of its low frequency, infrasound undergoes very little attenuation as it moves away from its source. In practical terms, this means it can be transmitted through forest and savannah for a considerable distance with little reduction in intensity. Sound emitted at a frequency of 1 000 Hz will decrease in intensity by a factor of 15 decibels for every 100 metres it travels through forest, and by a factor of 23 decibels through shrubbery or thick grass. However, in the case of infrasound, with a frequency of less than 30 Hz there is no discernable reduction in intensity in either of these situations. This raises the prospect of elephants exploiting the properties of low-frequency sound for the purposes of long distance communication.

To investigate this further, Payne shifted her centre of operations to the Etosha National Park in Namibia, where she and her co-workers built a six-metre observation tower close to a waterhole. Four microphones were set out in the bush, far enough apart for observers to identify which elephants were responsible for any given call. As Martin had noted in Zimbabwe, separate family groups frequently arrived at the waterhole within minutes of each other, even though no elephants had visited the site for several days. Large groups were to be seen suddenly to cease all activity and 'freeze' in response to an unknown stimulus (invisible and inaudible to the researchers), while at other times there was wholesale flight from the waterhole where peace and tranquillity had reigned only moments before.

When the real business of experimental science got underway, tape recorded elephant vocalisations were played back from a van two kilometres or more away from the waterhole while the behaviour of those animals around the hole was observed and filmed. Both males and females were found to respond instantly to a variety of distant playback calls, either by making calls of their own or, in the case of a male responding to an oestrus call, moving towards the source of the sound – arriving at the van several minutes later. Reactions were evident even when the source of the sound was moved to a distance of five kilometres. The infrasonic calls produced by elephants are able to reach such a distance partly as a result of their low attenuation and partly because they are loud (though still inaudible to us). Most register between 75 and 105 decibels – the human equivalent of walking past a busy construction site or listening to the loudest rock band in the history of music.

Ultrasound communication

At the other end of the sound spectrum, high-frequency noise (ultrasound) is employed by animals such as bats and dolphins as a location device, and possibly, in dolphins at least, as a means of communication. Ultrasound is produced at a frequency of 25 000–150 000 Hz, well above the range of human hearing.

Bat navigation

As early as the 18th century it was known that bats could find their way around after being blinded, but only if their ears remained unblocked. The development of radar around the time of World War II led finally to the realisation that bats emitted pulses of high-frequency sound that were reflected by solid objects and returned to the bats' highly developed hearing system in the form of an echo. The time taken for the signal to return gives a measure of the distance to the object. Because the wavelength at this frequency is so short, the bat is capable of forming a very high-resolution acoustic image of small objects such as the insects upon which it preys. In some species this is accurate enough to permit a spatial discrimination of only 0.3 millimetres between two surfaces.

When cruising around in search of food, a bat generates sonic pulses at a rate of approximately 10 per minute, but this increases rapidly to 100 per minute if a potential food item is detected. Curiously, the bat is not immune to damage in its own ears caused by these high-frequency pulses and, during sonic emission, it contracts small muscles within the middle ear to dampen down the movement of its ear ossicles. Some prey species of moths have developed mechanisms to detect ultrasonic pulses emitted by bats, while others such as the dogbane tiger moth have evolved the means to produce sonic clicks of their own, which they emit when under attack, thereby distorting the returning echo of the bat to such an extent that it misses its target.

Dolphin communication

Many species of dolphin apply ultrasound in ways similar to those of bats, as a means of procuring food and to examine their surroundings – a very useful tool in conditions where visibility in water may be limited. In addition, certain sonic clicks they emit may act as a type of communication. Bottlenose dolphins produce pure tone sounds that are unique to each dolphin (as are fingerprints to humans), and serve to identify a particular individual; they are known as 'signature whistles'. Since dolphins are gregarious by nature, these whistles may function to inform a group of who is present and who is absent, a useful adaptation when the turbidity of water precludes direct visual recognition. An odd and unexplained facet of their behaviour is the ability of dolphins to imitate the signature whistles of others. Perhaps this is a way of attracting the attention of the real owner of the whistle, especially if the imitator is in some sort of trouble and wishes to summon assistance from an older, more experienced, companion.

Body language

The tango emerged from the slums of Buenos Aires in 1880 as an amalgam of Spanish tango and milonga, a fast, sensual and disreputable local Argentinean dance. Originally possessing a lively tempo, it gradually took on a more melancholy pace, but anyone who has ever observed it cannot doubt it is a dance of intense passion and sensuality. Contrast this with classical ballet which developed out of the court spectacles in the Renaissance and later the French Ballet de Cour, in which social dances performed by royalty and the aristocracy were presented in harmony to music and pageantry. Training in the 17th century world of ballet was designed to mirror the deportment and manners of the nobility.

Whatever the format, dance communicates ideas and emotions. Aristotle felt that it 'represents men's characters and what they do suffer'. From the very beginnings of human culture, dance has played an important role; in more primitive times it served to bring the tribal group together as a means of preserving cultural identity, to augur success in the hunt and in war, to propitiate the gods for perceived infractions, as a celebration of the rights of initiation and fertility and to project the myths and religious convictions of the tribe. Despite the passage of time there has been remarkably little movement away from these basic tenets.

As a portrayal and communication of emotion, dance is uniquely human. But the use of body posture, movement and facial expression is widespread throughout the animal kingdom, conveying information as varied as the location of food, social dominance and submission, anxiety, affection, suitability for mate selection and the establishment of social bonds. These things are expressed in a variety of ways, depending on species, and it would require a much larger book to cover them all. Better therefore to consider a few examples that provide ample illustration of how these facets of behaviour are employed.

The dance of the honeybee

It took 20 years of meticulous study before Karl von Frisch unravelled the dance of the honeybee and, despite being awarded the Noble Prize for his efforts, he still could not quite believe the truth of what had been discovered. Small wonder, when the complexity of the dance language – involving, as it does, the mathematical ability to calibrate direction using a sun compass, together with a measure of distance and provision of information regarding the quality of a food source – is concentrated in a brain measuring less than one millimetre in diameter. By paint-marking bees that were foraging at solutions of sugar placed at various points and distances from the hive, von Frisch was able to observe their behaviour on returning to specially constructed glass-sided hives.

If the food source is less than 50 metres or so from the hive, the returning forager executes on the honeycomb a circular dance, which has been termed the 'round dance'. The bee moves rapidly in small circles, approximately her own body length in diameter, first to the left and then to the right. She is accompanied during this manoeuvre by other workers already on the comb, which follow and imitate her movements before flying off to gather food for themselves. The type of food is indicated by scent from the source adhering to the body of the bee.

The pattern of the dance changes when the food source is more than

149

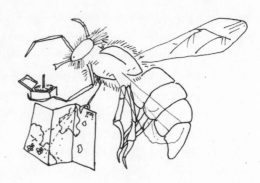

The dance language involves the mathematical ability to calibrate direction using a sun compass, together with distance and the quality of the food source.

100 metres distant. Instead of describing a purely circular pathway, the bee now adds a short, straight run during which she vigorously wags her abdomen from side to side – the 'waggle dance'. What is truly remarkable about this dance is the way it indicates direction. The convention among honeybees appears to be that the position of the sun at any one time of the day is to be represented by the top of the comb. If the food source is located directly towards the sun, the straight part of the waggle dance is directed vertically upwards, while a waggle dance directed vertically downwards indicates a food source directly away from the sun. Between these extremes, the dance is angled on the surface of the honeycomb to correspond to the angle of the food source relative to the position of the sun. Distance from the hive is indicated by the speed of the dance: the faster the dance, the closer the food source. As with the round dance, fellow workers follow and imitate the dance routine of the returning forager. On cloudy days, the position of the sun is replaced as a reference point by a patch of polarised light in the sky.

Although von Frisch observed these dances through glass-sided hives, it is worth remembering that, under normal conditions, they are performed in complete darkness. Their information therefore cannot be communicated by vision and must involve a perception of the dance's movement through direct contact with the legs and antennae of the surrounding bees.

Since the initial landmark studies on bee communication, further information has come to light in some aspects of the waggle dance. The quality of a food source, as measured by the sugar content of nectar, is reflected in the degree of enthusiasm exhibited by workers on the honeycomb when a forager returns with her load. Food from a rich source is unloaded quickly by the workers, usually in 30 seconds or so, and this elicits in the forager a corresponding enthusiasm in her subsequent dance routine. A poor quality source on the other hand may either be unloaded slowly or not at all, and is followed by a lacklustre performance in the dance, which is usually ignored by other bees.

But is it language?

The school of science known as behaviourism believes, and possibly always will, that no species other than our own has the capacity for conscious thought and reflection, with the possible exception of other primates. Thus, most forms of life are reduced to a host of automatons which 'nourish a blind life within the brain'. But if the sounds made by elephants do constitute a form of language, there seems little choice but to accept the unpalatable truth of this being a species capable of conscious thought. Language requires thought, for what is language if it is not a means of conveying to another of the same species some form of meaningful idea or instruction. In her book 'Coming of Age with Elephants', Joyce Poole quotes the definition of a 'word' as developed by scientists studying the language capabilities of non-human primates. According to this, to qualify as a word, a communicating signal must (a) be an arbitrary signal that stands for some object, activity or relationship; (b) contain stored knowledge; (c) be used intentionally to convey this knowledge, and (d) be capable of eliciting an appropriate response in the recipient. The rumblings of elephants fulfil all these criteria, but rather than words they more closely resemble phrases and ready-formed ideas.

Joyce Poole firmly believes this is language – but she can never quite bring herself, at least in public, to the conclusion that elephants possess consciousness and self-awareness. However, the use of language, even a fairly rudimentary one, leads to the inescapable conclusion that an idea has been formulated, transmitted and understood. How can this proceed without conscious awareness?

CHAPTER TWELVE

A STORY OF GREED

the future of the
African elephant

*Elephant numbers have increased to a level where many
experts feel they will severely degrade their habitat.*

Continent of extremes

Africa is a continent of extremes where it is possible to be
overwhelmed by the beauty of a place one day only to be confronted
with horror and despair the next. As a doctor in rural Africa I was
faced with this paradox on a daily basis. It is a land where indolence
and poverty are the norm and where corruption steals the will of
even the strongest. Into this mix is thrown some of the world's most
exciting and exotic wildlife, whose existence remains tenuous in the
face of human population growth. Fortunately, African governments
believe wildlife to be a 'good thing' in general, if for no other reason
than that it equates with tourism, which in turn equates with money.
For the people of the land however, who are forced by poverty and
bureaucratic indifference to exist on the basis of a local subsistence

economy, wildlife is at best an irritation simply occupying land better suited to livestock or the plough. In the uneasy zone between wildlife sanctuary and farmland, conflict is inevitable – crops are plundered, cattle eaten and wild animals shot or poisoned. For better or worse, this resentment has polarised, to a degree, around the world's largest land mammal – the African elephant – known not only for its prodigious appetite and capacity for destruction but also for the monetary value of its body parts.

Elephants in general, and particularly the African elephant, have fared badly at the hands of humans over many centuries. They have been enslaved into functioning as beasts of burden, modes of transport, instruments of war and circus entertainers, while on a more sinister level they have been dismembered into a variety of trinkets and chattels in the form of jewellery, oriental carvings, exotic inlays, piano keys, billiard balls, footstools, umbrella stands, wastepaper baskets and flywhisks. Unlike the mountain gorilla, they have not yet been converted into ashtrays. Protection for the elephant has arrived very late in the story but, with dreadful irony, it is this very protection that may ultimately pose a greater threat to the elephant in the modern era. Confined, by and large, within the strict limits of national parks and reserves, elephant numbers have increased to a level where many experts feel they will severely degrade their habitat such that it not only threatens their own future, but also that of a large number of other plants and animals.

Over many years there have been discussions on the management of elephants, usually on an ad hoc basis with little in the way of consensus emerging. But many feel that now is the time for serious debate (some would say it is already too late) because the issue extends beyond the elephant to a consideration of the ecosystem as a whole, and whether its diversity can remain intact. A unified 'pan-African' approach will be difficult to achieve, given the large variations in the distribution of elephants across the continent. In East Africa there is the view that elephant numbers have yet to recover fully, following the orgy of poaching in the 1970s and 1980s, which may have been responsible for the loss of as many as 80 per cent of the elephant population. On the other hand, elephant numbers in southern Africa are reported as being excessive, particularly in Zimbabwe and Botswana, where many believe the habitat is already beyond saturation point. Perhaps not surprisingly divisions are apparent between conservationists, legislative bodies, government departments, scientists and animal rights organisations, and the passion with which opinions are expressed gives a good indication that no easy answers are likely to be forthcoming. The discussion that follows does not support any particular point of view, but tries to represent a spectrum of facts and the views of others in an objective way, so the reader can formulate his or her own judgements.

Natural history of the elephant

Everything about the elephant is large. A mature African bull elephant stands four metres in height, weighs five to seven tonnes and daily consumes 300 kilograms of woody plants and grass, drinks 200 litres of water and produces 150 kilograms of dung. Elephants live long lives, comparable to humans, and in the absence of predation, disease or poaching may easily attain the age of 60 years or more. Natural life span is determined by the peculiar dental arrangement in which six sets of molars sequentially erupt throughout life; when the last of these become sufficiently worn, death from malnutrition is inevitable (see Chapter 2). Females are somewhat smaller than males and become capable of bearing offspring at the age of 10 to 11 years. Intervals between calvings vary between four and eight years and the net population grows (by compound interest) at a rate of four to six per cent per annum. Numbers therefore effectively double every 12 to 15 years, an important observation when elephants are restricted in their distribution to the defined areas of national parks.

The elephant as commodity

Elephants have been wandering over the African landscape for at least 800 000 years and, for most of that time, have enjoyed unlimited access to the habitat of their choice and a seemingly infinite food supply. But this pachyderm utopia could not hope to last forever and was brought to an end by a growing demand for ivory and the exponential growth of the continent's human population.

From the time of the Pharaohs some 4 000 to 5 000 years ago, ivory was a prized possession in Mediterranean countries, a status symbol of wealth, power and influence. Carved and rendered to luxury goods such as jewellery, sculpture and elaborate decoration, the demand was ever present. King Solomon was said to have constructed his throne from

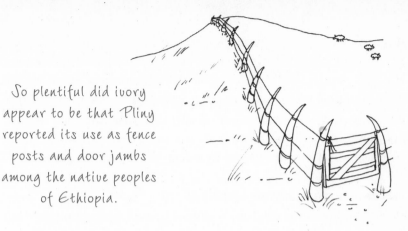

So plentiful did ivory appear to be that Pliny reported its use as fence posts and door jambs among the native peoples of Ethiopia.

ivory and the palace of King Ahaab was so heavily decorated with it that it became known as the 'Ivory house'. So plentiful did ivory appear to be that Pliny reported its use as fence posts and door jambs among the native peoples of Ethiopia. Thus far, Somalia had been the major source of ivory, shipped up the Red Sea to Egypt, but the arrival of the Portuguese on the west coast of Africa in the 15th century, and later in Mozambique, allowed Europe to have direct access to this invaluable resource. For the elephant, it has been a downhill story since then. Their only defences were the inhospitable nature of the African interior combined with a motley army of creatures against which *humans* had no defences – mosquitoes (malaria), tsetse flies (sleeping sickness), a microscopic snail (schistosomiasis) and other sundry parasites. These were to prove sadly inadequate against the firearms placed in the hands of colonial representatives, opportunists and a ragbag of European misfits. In his book *Africa's Elephant*, Martin Meredith reports that during the glory years of colonial occupation, Great Britain alone was importing an average of 500 tons of ivory each year. This went on for 60 years or so and was part of a worldwide trade, which annually cost the lives of approximately 65 000 elephants. How close the elephant came to extinction during this period is unknown, but the carnage became sufficiently large finally to provoke alarm among colonial governments. The end result was to restrict hunting quotas and create protected areas in the form of national parks and reserves.

Quiescent for many years, the ivory trade underwent a further renaissance when the economic boom in Japan and other Asian countries fuelled an increased demand for ivory. The price of ivory leapt from $US5.50 per kilogram in the 1960s to $US300 in 1989; and this, combined with the ready availability of automatic weapons, set the stage for further slaughter. With these new weapons the elephant population was decimated in a way that had not been possible before, and the process was facilitated by the corruption and bribery endemic in Africa, including the involvement of officials charged with the management and protection of wildlife. Over a 20-year period spanning the 1970s and 1980s, the total number of elephants across the African continent was reduced from 1.3 million to 600 000. The poaching gangs had primarily targeted the East African nations of Kenya, Tanzania and Uganda, as well as the Congo Basin and Zambia, leaving the populations of Botswana, South Africa and Zimbabwe relatively intact – a discrepancy which, as we shall see, formed the nucleus of bitter divisions of opinion in later years.

Responses to trade in elephants

The feeble response of the international community in 1976 was to place the African elephant onto the Appendix II list of endangered species as drawn up by the newly created organisation CITES (Convention on International Trade in Endangered Species of Flora and Fauna). Given the levels of government corruption, this measure had absolutely no effect in curbing poaching activities because all it required was for the export of ivory to be accompanied by an export permit, easily obtainable in the circumstances. This situation remained unchanged for a further 10 years, at which time CITES recommended that each ivory-exporting nation was to restrict its export of ivory to a fixed quota. This legislation led to some ludicrous situations in which ivory was 'diverted' for export into countries with very few elephants but which were nevertheless granted substantial quotas. Thus Burundi, which had killed all but one of its elephants (see Jeremy Gavron's book – *The Last Elephant*) found itself exporting 1 300 tons of ivory between 1965 and 1986, all smuggled in from the neighbouring countries of Tanzania, Congo and Zambia. Although duly castigated for its behaviour and with promises to reform, Burundi still exported a further 200 tons of ivory over the next two years.

The legislation enacted by CITES only covered the export of raw ivory, there being no limit to the movement of partially or fully worked tusks. Minimal processing thus allowed trade to continue unhindered without the need for licences and led to the construction of factories, particularly in the free trade zones of the United Arab Emirates, which either cut up or lightly carved tusks for legal, ongoing export to Hong Kong, China and Japan. This story was eventually exposed in 1988 by members of the Environmental Investigation Agency – a wildlife activist organisation based in England, who were successful in filming such an operation in Dubai and Hong Kong.

One further decision made by CITES was destined to provide another little flurry in the ivory trade. In attempts to garner more support and increase its membership, CITES announced that all stockpiles of tusks in the possession of non-member states could now be legally sold. This released a massive amount of ivory onto the world market with handsome profits for all concerned. The trade in ivory had never looked better. Around this time the voices of elephant experts and scientists were beginning to be heard who, far from expressing dismay at the plight of the elephant, were advocating the controlled killing (culling) of elephants on the basis that elephant numbers were declining, not as a result of unrestricted poaching, but because their numbers exceeded the capacity of the land to feed them. Controlled killing, it was argued, would maintain an appropriate balance between elephants and nature, with the added benefit of utilising the culled elephants as a resource material, providing meat for protein-starved local populations and export-quality ivory to give much needed currency to impoverished African states. This philosophy gained considerable support among the southern African nations of Zimbabwe and South Africa, which still held large elephant herds and had not suffered as acutely from the poaching that had ravaged East Africa. Nevertheless, a report commissioned by CITES (bizarrely using funds donated by the Kowloon and Hong Kong Ivory Manufacturers' Association) to investigate the true status of elephant numbers revealed an alarming picture in which elephant populations were predicted to become 'commercially extinct' over the whole continent within 25 years. The report was never, officially, to see the light of day. The views it expressed, however, were supported independently by the Ivory Trade Review Group at a meeting in London in 1989. To maintain ivory supplies at a constant level required more and more elephants to be killed – once the 'big tuskers' had gone, poachers were driven to take smaller tusks from smaller animals with a substantial increase in elephant mortality.

Much of this story unfolded out of the public eye, and the ivory trade, if it was thought of at all, was considered generally to be an acceptable enterprise. With the usual mental sleight of hand peculiar to humans, there seemed to be little consciousness of a connection between ivory trinkets, expensive inlays and dead elephants. It is therefore a testament to the tenacity of a rare group of people who finally exposed what was really happening in Africa, and the true nature of the ivory trade. Only the groundswell of public opinion finally overcame bureaucratic inertia, forcing on CITES the decision to include the elephant on Appendix I, prohibiting international trade in ivory. Poachers were thereafter discouraged by a 'shoot-to-kill policy' developed by many governmental wildlife departments.

The ivory trade is a sordid story and is difficult to present in any other light: a story of greed, venality, indifference, bureaucratic incompetence and political expediency. But of course it will never entirely cease as long as Asian countries create a demand and their peoples are willing to purchase ivory products. Not all members of CITES were in favour of imposing a ban on the trade of ivory, particularly vigorous objections coming from southern African countries. Rightly or wrongly, the belief that elephants are in the process of destroying their own habitat to the detriment of themselves and other species is gaining momentum, especially so when these animals are trapped within the finite boundaries of national parks and game reserves with their limited natural resources. This argument needs to be examined in some detail, particularly since it is currently being used to lobby CITES into reconsidering the status of the elephant on Appendix I, with a view to downgrading it once again to Appendix II. The supporters of such action have been quick to point out that their position is determined by the desire to save whole ecosystems from destruction, to protect the elephant, as it were, from the follies of its own actions. Despite the lessons of the past, nobody is talking about the likely resurgence of the ivory trade and the behaviour associated with it. But just how big a problem is there in regard to elephant populations in national parks? Is culling the only option available? Should ivory resulting from elephant culls be burned and the ban on ivory trade remain in place? Where is the money from ivory likely to end up?

How many elephants are too many?

The rapid expansion of the human population in Africa has ensured that the once pristine areas of wilderness set aside for both flora and fauna have, in modern times, become islands in a sea of human settlement. Conflict between humans and animals at park boundaries is rife and is fuelled, in many cases, by resentment of the present generation towards wildlife occupying land that was home to their ancestors, and to which they are now denied access. Realistically, human population pressure is such that there is little, if any, chance of setting aside new areas of wilderness as wildlife sanctuaries. The only exception has been the recent proposal to amalgamate South Africa's Kruger National Park with the Gonarezhou Game Reserve in Zimbabwe and the adjacent areas of Mozambique to form a 'megapark'.

A fundamental purpose behind the creation of national parks was the desire to encourage and maintain a high degree of diversity in both plants and animals, free from human interference. Unfortunately, elephants, like humans, have no concept of the finite nature of resources. The problem with elephants is not so much their appetite, but their capacity for destruction in terms of knocking down mature trees and killing others by stripping the bark. Forest can regenerate, of course, but this takes a long time – too long, in fact, given repeated assaults by elephants and other browsers. One of the many secondary culprits here is the impala, with its preference for the fresh new leaves of regenerating saplings. The end result of these activities is progressive deforestation and conversion of the landscape to open savannah and grassland. This may be good news for grazing animals, but the downside is the loss of diversity and the irreversible loss of some species altogether.

Some argue that change in any habitat is inevitable, that nothing in nature stands still and that it is the way of evolution to adapt and change over time. True enough; but this belief carries much less conviction when applied to circumstances pertaining in areas with artificially fixed boundaries. Within the parks we are not witnessing wild nature in action, but a man-made spectacle where restriction of movement and resources modifies behaviour patterns. In a utopia of infinite resources and space, there would be substantially less impact on the environment; but much as we may prefer otherwise, the modern reality is of flora and fauna congregated within the relatively narrow limits of regions whose boundaries are determined by humans. The

... trees also play an essential role in physically holding the land together.

growth rate of human populations and the animal populations within reserves cannot help but put stress on the whole landscape.

The death of mature trees has more far-reaching effects on the land than on the unwelcome exposure of shade-loving plants and forest-dwelling creatures, because trees also play an essential role in physically holding the land together. Without the earth-binding influence of tree roots, tropical rains remove the nutrient topsoil, producing huge erosion gullies and leaving behind ground that no longer has the capacity to nurture any but the most hardy plants. The results may be a lunar landscape with a permanent loss of most species. Added to this are the more natural phenomena that alter the structure of the land in the form of drought, fire and tectonic upheavals.

One of the problems in considering the impact of one species or another on the landscape is the almost complete lack of hard scientific data concerning the natural history of the vegetation itself, as well as the interrelationships between different species of plant that make up the whole. The plant composition of a given habitat is likely to change as much by cycles of drought and fire as it is by predation, so great care has to be taken in attributing a perceived degradation of habitat to the actions of a single species of animal. It is also worth bearing in mind that,

160

although elephants are destructive, they are essential to the reproduction of many plants (see Chapter 4). The seeds of several West African trees will not germinate unless they first pass through the gut of the elephant.

Evolution tells us that populations of animals wax and wane according to prevailing conditions, a bountiful food supply resulting in a peaking of population only to be followed by a crash as conditions change and food supplies diminish. With perhaps the exception of Tsavo National Park in Kenya, there has been no obvious sudden natural reduction in elephant numbers, though this may well have been obscured by the activity of poachers. Wildlife enthusiasts have pointed to the possibility that elephant numbers may follow similar trends with peaks and troughs, but operating on cycles measured in periods of many years, perhaps as many as 100 years or more, and that what we are witnessing at the present moment is part of the growth phase. But this would imply a remarkably consistent population trend across the whole continent, which is unlikely, given the variable nature of habitat; and there is no scientific evidence to confirm or refute this hypothesis one way or the other. So, for now, it must remain a theoretical possibility but no more.

From what has been said already it is clear there are a number of unknowns in regard to the biology and ecology of the plants and animals found in Africa's wild places. The sight of broken trees should not result in a knee-jerk response to bring out the guns, but nor should any possible remediable action be delayed until once beautiful woodland is on the verge of desertification. If the purpose of national parks is to promote diversity, then action should be based on evidence supporting a definite reduction in such diversity. If cycles of vegetation are not understood, the task is that much more difficult – but it has to start somewhere. What will be used as the 'gold standard' to determine when a habitat is under serious threat? This should be the province of skilled biologists, but thus far they have had little to contribute outside the sphere of specialist journals. This is an important question because it sets the whole tone of wildlife management policies within these areas. Decisions cannot be left to those who shout the loudest or who have a hidden agenda.

Changes in the dynamics of animal populations within reserves and their effect on local habitats have placed park authorities and governments in an invidious position. If the diversity of species is to be assured, active management strategies become unavoidable and bring

161

a tremendous responsibility which cannot be shirked. Nobody envies the park authorities in their task, battered as they are from all sides with every conceivable shade of opinion.

The culling option

A culling operation involves rounding up whole nuclear families – matriarch plus her offspring, siblings, nephews and nieces, aunts and perhaps her mother – and then shooting them. This is repeated again and again until the numbers have been judged sufficient. Between 1968 and the mid 1990s it was policy in South Africa's Kruger Park to cull between 400 and 500 elephants each year to maintain a stable population of about 7 000 animals. Culling elsewhere has been more sporadic, such as the killing of 2 000 elephants over a two-year period in the Murchison Falls National Park in Uganda, and 3 000 elephants over a three-year period in Zimbabwe's Hwange National Park. Abhorrent though most people find this practise, there is some biological sense to it when the habitat shows an unmistakable trend towards serious and perhaps irreversible damage. But, with lesser degrees of damage, are other alternatives feasible?

Destruction of mature woodland is largely inflicted by bull elephants and only occasionally by matriarchal family groups. If culling is deemed to be necessary, then why not selectively target the bulls, which would not only preserve habitat but also decrease reproductive potential, and so help to keep numbers within manageable limits without the need for wholesale slaughter? As far as I am aware, this proposal has never been subjected to scientific investigation.

The culling of elephants has polarised opinion among wildlife authorities, biologists and animal rights groups like no other issue. Most accept the principle of such a strategy when no alternative exists to save the environment. But culling also raises the issue of commerce – sustainable utilisation, which increases the temperature of the debate considerably. On a continent where protein calorie malnutrition is almost universal, it would be stupid and inhuman to deny the population access to the meat derived from culling operations. The same, however, cannot be said for ivory. The African states of South Africa, Botswana and Zimbabwe, all with large animal populations, have lobbied CITES to exclude them from the restrictions imposed by the ivory trading ban. With proper control, they argue, ivory should be

Culling involves rounding up whole nuclear families — matriarch plus her offspring, siblings, nephews and nieces, aunts and perhaps her mother.

made available on the world market, in limited amounts, in a way that would satisfy the customers and represent significant export earnings for the countries involved. This may seem quite feasible in theory but, given the past history of the ivory trade, there is a well justified fear that it may lead to the re-establishment of a parallel, illegal ivory market and all the activities associated with it. In this light, sustainable use has to be viewed with a degree of scepticism because it is so open to abuse. The necessity of culling what may be a substantial number of animals needs to be formulated on sound scientific evidence designed to protect a deteriorating habitat, and not with an eye to the commercial benefits which may accrue.

Elephant contraception?

Contraception is not a word that readily springs to mind in connection with elephants, but this may be one of the keys to limiting numbers in the long term. Initial attempts to control fertility using oestrogen implants in females were disappointing, leading to unusual behaviour patterns including moving away from the family group and abandoning offspring. The high dose of oestrogen also induced a state of false oestrus, resulting in females being continually pursued and harassed by males intent on copulation. A more promising approach has been developed by Richard Fayrer-Hosken and his co-workers at the University of Georgia, who have pioneered the use of a contraceptive vaccine. Derived from the protein coat surrounding the ovum of the domestic pig, pZP vaccine (pig zona pellucida) when injected into female elephants triggers the formation of antibodies, which create a protective barrier around the elephant ova, thus preventing penetration by sperm and hence fertilisation.

Female zoo elephants treated with pZP all developed antibodies, which persisted for 12 to 14 months. Subsequent field trials were conducted in the Kruger National Park, where 21 elephants were vaccinated and given booster doses at six weeks and six months later. Twenty further elephants were injected with inactive placebo doses, and served as controls. One year later the incidence of pregnancy in the control group was 89 per cent compared to 44 per cent in vaccinated animals. As a consequence of this information, the vaccination schedule was changed to administering booster doses two and four weeks after the initial dose. With this new regime, pregnancy rates fell to 20 per cent. Over a period of one year, no abnormal behaviour patterns emerged in treated animals and subsequent pregnancies proceeded normally, thus showing the effects of the vaccine to be readily reversible.

Extending this work to large herds of free-ranging elephants poses major logistical problems, particularly when an individual

Initial attempts to control fertility using oestrogen implants in females were disappointing.

elephant requires to be identified accurately and given three separate inoculations, to be repeated a year later. These are early days but, for now, this technique may be better suited to small populations. Some authorities have also urged caution with the use of immuno-contraception, believing it may adversely affect the ability of the immune system to react adequately to the challenge of a variety of infectious diseases, though proof of this is not yet to hand. Clearly, further small-scale trials with prolonged follow-up need to be under-taken before vaccines of this type go into general use. Translocation of elephants from one area to another represents a further alternative to avoid overcrowding, but becomes too expensive and impractical unless numbers are small and distances relatively short.

Priorities for Africa

The West has not been slow in expressing criticism of African wildlife management policies but, on a continent consumed by poverty and corruption, sinking under the burden of unpayable debt, African governments are concerned with greater priorities than wildlife. With whole populations living below subsistence levels, the irony is not lost on Africans when strident demands are made upon them to provide adequate protection for wildlife, particularly when such demands come from former colonial powers that have plundered Africa's natural resources of oil and minerals and indeed continue to do so through controlling shareholdings in present-day companies. In the economic circumstances confronted by Africa, its peoples believe wildlife to be a legitimate resource to furnish much needed income, whether this is derived from tourism, hunting or the sale of animal products from culling operations. And this claim becomes increasingly difficult to deny when the World Trade Organisation and the governments of G8 countries continue to manipulate prices and access to markets in favour of developed nations. With justification, former President Masire of Botswana has termed this western attitude 'environmental imperialism'.

If the rest of the world wishes to preserve wildlife on the African continent and see it flourish, then it will have to dig into its pockets or, at the very least, allow African goods an equitable share of the world's markets. The elephant is a symbol of the fragility of all wildlife on the African continent and its survival will not depend on sentimentality or goodwill, but on raw economic realities.

CHAPTER THIRTEEN

WALKING THE WALK
social dynamics

Dominance among monkeys and apes is not difficult to spot, being worn like a badge and emphasised by body posture and behaviour.

Social dominance and submission

In most animals there is a structural hierarchy where each individual knows his or her own place in the social milieu, otherwise known as 'the pecking order', and which is determined largely by sex, age, size, health and the level of aggression innate in any one individual. Dominance among monkeys and apes is not difficult to spot, being worn like a badge and emphasised by body posture and behaviour. Typically, a high-ranking male displays his confidence with a swaggering gait, taking in his surroundings with a casual eye and frequently walking directly towards a lower-ranking individual with the sole purpose of displacing the subordinate from whatever activity he happens to be engaged in. A dominant animal may choose at times to remain more aloof from the group, adopting the role of sentinel, in which he finds a vantage point and faces away from, and largely ignores, his fellows. This is often accompanied by behaviour referred to as 'the penile display', sitting with legs apart to expose a partially erect penis – a form of behaviour particularly common in species with bright genital colouring, such as the vervet monkey, which possesses a bright blue scrotum.

As a child I was taught to avoid staring at people, even other children, since this was not only considered rude but likely to evoke discomfort, irritation and possibly hostility. Not surprisingly really, because among the rest of the animal kingdom, staring constitutes a challenge or a direct threat. It is employed by monkeys and carnivores alike and often occurs in association with yawning – not, in this case, to express boredom or fatigue but to display and advertise canine teeth and, by implication, the likely consequences of returning a challenge. Unless there is a real and unequivocal challenge to take over the control of a group, these displays seldom result in actual fighting, both animals being likely to sustain severe wounds from such powerful teeth. (This does not prevent teeth from being used freely on an obvious subordinate should he or she transgress acceptable social boundaries.) Threats directed towards predators or a rival group of monkeys involve a great deal more in the way of dynamic behaviour than that usually seen in sorting out the hierarchy in a given group. Here, there is more likely to be a lot of violent movement in the trees, with much branch breaking, branch shaking, jumping up and down, loud calling, slapping the ground, throwing objects and head bobbing, together with the usual staring and display of teeth.

As with aggression, submissive behaviour in monkeys has a number of different facets. The simple expediency of escape from a dominant animal is less rewarding than one might imagine, because the individual is then essentially forced to run a gauntlet where admonition may come from any number of animals superior in rank to the offender. The easiest act of appeasement is basically to ignore, as far as is possible, the attentions of a more dominant animal. This can be achieved by pretending indifference, appearing overly preoccupied with a given activity and electing not to notice an intense stare. A more nervous response may come in the form of looking away (visual cut-off), avoiding the gaze of an opponent – the simian equivalent of a child believing itself to be invisible by placing its hands over its eyes (usually peeking between gaps in the fingers). Other facial expressions denoting submission include grimacing – with lips retracted and teeth clenched – teeth chattering and lip-smacking. In both sexes, submission may be expressed by adopting the position usually exhibited by the female during copulation. This may be so important and evolved in some species that males are able to mimic the discoloration of the skin surrounding the genital area normally seen only in oestrus females.

Aggression – do we need it?

Aggression is almost universal throughout the animal kingdom. Konrad Lorenz, Nobel Prize Winner for his work on animal behaviour, believed aggression to be inevitable, an urge like hunger or sex which can only be dissipated by performing the appropriate act. On this basis, aggressive tendencies were perceived as gradually building to a point where overt aggression explodes and, having been discharged, is followed by a variable period of quiescence. In this view, aggression becomes a purely innate, genetically inherited mode of behaviour, largely beyond the control of the individual. This implies a rigid behaviour pattern, which may not always be appropriate depending on the particular circumstances.

Direct observations in the field and laboratory do not support these conclusions. Rather than exhausting itself when its objective is secured, aggression shows a marked tendency to persist for a substantial time after the event and, in some cases, to escalate. According to the model proposed by Lorenz, an animal isolated from its fellows should become more aggressive as time passes because it lacks a focal point for its aggressive behaviour. In fact the reverse is more often true, aggression diminishing with time precisely because the potential reasons for conflict are absent.

A further problem arising from Lorenz's theory is that it does not take into consideration the role of experience. From an evolutionary standpoint, aggression certainly serves as a means through which animals compete for limited resources, be this food, territory or access to females. But outright combat on all occasions does not serve Nature's purpose, resulting as it might in the field being strewn with dead or severely injured protagonists. Much better, then, for all concerned to keep fighting to a minimum and employ less violent strategies to communicate their entitlements.

Discretion versus valour

An animal can effectively gauge its chances of success by careful observation of its opponent, using parameters such as size, muscular development, body posture and others, including vocal cues and specific secondary sexual characteristics. A good example is provided by the 'roaring contests' between red deer stags in the rutting season. The dominant stag controls a harem of females but, as the rut advances, he is required to fend off a number of challenges to his ownership. Instead of engaging in endless physical battles, matters are resolved for most part by vocal contests. From a suitable distance, the incumbent and challenger roar at each other, dominance being determined eventually by whichever can roar at the faster rate. Presumably this is taken as a statement of physical fitness or stamina and an indirect way of expressing likely fighting ability. If roaring does not show a clear winner, the two animals then join in a parallel walk, which allows each of them to assess the physical attributes of the other. And if this fails to deter the challenge, a fight becomes inevitable.

Controlled aggression such as this is not wholly innate and is based partly on learning as a result of previous experience. In this way, a hopelessly inadequate male need not pay the price of injury or death following an ill-judged attempt to challenge a clearly superior adversary. All in all, aggression is an integral part of life, often the only way an animal can gain access to a limited resource. But, over a long period of time, this has become tempered to a more or less ritualised display, especially in animals where serious injury can be anticipated in the event of a full-blown physical confrontation; a lesson, one might add, that we humans as a species have singularly failed to grasp.

Red deer stags engage in 'roaring contests' in the rutting season.

Establishing social bonds

At some time in their adult lives all mammals find it necessary to come together in some form of grouping. This may, like the solitary cats, be a very brief encounter simply for the purpose of breeding, while others prefer to congregate in their hundreds or thousands for prolonged periods. There are distinct advantages in group living: it provides a better warning and response system to the approach of predators, and it allows a number of animals to utilise a food source discovered by one or two of their fellows. But living as a group does not necessarily mean that social interaction occurs between its members. There is no evidence, for example, of social bonds within a herd of ungulates, other than between mothers and their offspring. Partly, this is a reflection on the size of the group – you only have to cross the city of London on the underground at nine o'clock in the morning to be aware of your own total anonymity and the indifference of fellow travellers to your existence. Formation of close social bonds is largely limited to animals that restrict the size of the core group. Contact within the group may take a number of forms, but the formation of social bonds seems to occur almost exclusively by touch – grooming, rubbing, licking and playing.

Gaining trust

The vocalisations made by primates during grooming are, in the opinion of some authors, the forerunner of an oral language that only finally emerged in our own species. These noises are said to be the functional equivalent of gossip, thus anticipating the arrival of the ladies' hair salon by several million years. Be that as it may, grooming is certainly an important part of the glue that holds primate society together, and may occupy 10 per cent or more of the time of an individual. Though seen more frequently between mothers and offspring or siblings, grooming provides a means by which subordinates can approach more dominant animals without incurring an aggressive response. Grooming is obviously a pleasurable activity, at least for the recipient, and there is some work suggesting that endorphins (pleasure hormones) are released during the process.

Lacking the dexterity of primates, the cat family grooms by licking. Seeing two lionesses of the same pride meeting after a few days apart, it would be churlish to describe their greetings as anything but affectionate – rubbing their heads together with eyes closed followed by a good deal of head licking. The grooming of cubs also proceeds with vigorous licking. Regardless of the mechanics adopted by each species, grooming seems to acquaint one individual with another and, because it involves direct tactile body contact, it is likely to induce a sense of trust, which might otherwise be absent in a purely vocal communication.

Play as a precursor to prey

Play behaviour is common among mammals and even birds, but is particularly prevalent among primates and those carnivores that form social groups. Young carnivores indulge in play that mimics the actions and strategies they will later employ as predators to capture and subdue prey – a clever gambit on the part of nature to combine fun with education. Because this behaviour has the potential to cause significant injury to the participants, some understanding of the benign nature of the game needs to be established between them. These play signals may be expressed through body language and American evolutionary biologist Marc Bekoff has described such behaviour in canids, which he terms a 'bow': crouching with the four limbs on the ground and the hind end of the animal remaining upright, usually to the accompaniment of barking and tail wagging – behaviour familiar to all dog owners. Play serves to establish social bonds, co-operation between group members and the physical fitness of the individual. Curiously, there also appears to be a sense of fair play, in which the 'winner' of various contests is distributed on a roughly equal basis – animals that 'cheat' or bully their playmates are frequently avoided. As with grooming, fairness may encourage the development of trust, a crucial factor among animals such as lions or wolves, which frequently prey on species much larger than themselves and where group co-operation, and trust in the actions of another, become essential to success.

Courtship rituals

Sit at an outdoor café during the summer months anywhere in Western Europe or South America and you will witness a parade of groups of young men and women sauntering along the boulevards. More than simply a fashion parade, this perambulation is a display, a demonstration of attractiveness, where participants can peruse and admire the opposite sex with the possibility of engaging a suitable partner. By no means original, humans are merely following a precedent established by other members of the animal kingdom over many aeons. Visual signals are important indicators in reproductive behaviour and are used by both sexes to assess the genetic quality of potential mates. A scruffy, malnourished, timid individual is not likely to score highly in the reproductive stakes.

Evaluating the talent

A lek is an area where certain species of birds or animals gather solely for the purposes of reproduction. No food resources are available and, typically, each male may occupy a small territory within the lek. Females arrive in numbers to evaluate critically the qualities of likely suitors, which goes a long way to explain why the males of these species are so elaborately adorned with striking markings and plumage, in marked contrast to the dowdiness of the females. Lekking is a feature of polygamous species, where the male functions simply as a sperm donor, playing no further part in the reproductive and nurturing process.

Precisely how a decision is reached favouring one male against another is not well understood, but one species, the black grouse, possesses the amazing ability to predict the longevity of a prospective male – steadfastly avoiding those it considers to have a limited prognosis and thereby reducing the risk of transmitting inferior genes to offspring. As females appear on the scene, the male grouse performs a display that highlights prominent features, such as his white tail and bright red comb and wattle. He may also engage in fighting with the male in the adjoining territory. Researchers studying the behaviour of these birds were unable to identify any single or group of physical characteristics that determined the females' choice. But further investigation some months later showed that the females had chosen mates which had survived the intervening period, whereas those they had rejected were dead. Reviewing the data, the researchers were still unable to pick out males that were likely to perish – assessing them in terms of parasite load, fat distribution, various body measurements or general appearance. Clearly, the females were able to recognise some subtle signal which spoke badly for future survival, but its nature remains unknown. Nevertheless, in some species, recognition of poor health by detection of a heavy parasite infestation may be possible, and raises the question of whether olfactory cues may be part of the mate selection process rather than purely visual ones.

The cost of being spectacular

Birds go to extraordinary lengths to attract females, artifices that employ brilliant plumage, gifts of food, nest building and elaborate dances or aerial routines. Perched on a thin branch over the water or among reed beds away from snakes, the golden weaver presents a beautiful, spherical nest to a prospective female. It takes him several hours to complete the work and construction is always under the threat of demolition by a rival male. If the female disapproves of the result, he will start the whole business again from scratch.

The lilac-breasted roller, the size of a European starling and one of Africa's most beautiful birds, takes its name from it courtship behaviour. In its attempts to attract females, it flies to a good height and then drops in complete freefall. Just as collision with the ground seems inevitable, it pulls out of the dive and, with a flash of brilliantly iridescent wings, executes an elaborate rolling manoeuvre after the fashion of an acrobatic air-show stunt.

Having found a suitable nesting site on some remote oceanic island, the male frigatebird performs a more sober, but physiologically demanding, display in which he inflates a huge, crimson-coloured throat pouch – to the point where explosive decompression seems to be ever imminent. Other birds such as peacocks, birds of paradise, lyre birds, whydahs and pheasants display their spectacular tails.

The male frigatebird inflates his throat pouch to the point where explosive decompression seems to be imminent.

But elaborate markings may come with a price tag. Avian females are dull and dowdy because it is in their evolutionary interest to be so, committed as they are to long periods of immobility while incubating eggs. In this circumstance, the absence of conspicuous plumage becomes a device to avoid the attentions of predators. In stark contrast, the brilliant coloration of the males acts as a magnet to predators, and the possession of a long, cumbersome tail may significantly hinder escape. Nevertheless, evolution has determined that the benefits accruing from this colourful structural extravaganza clearly outweigh the risks.

Communication: the rise and rise of deception

The advantages of communication are obvious. Communication within a species allows transmission of information from one member to another on matters as varied as food location, danger and sexual receptivity. In addition, it facilitates the formation and maintenance of social groups and establishes a social order. The information highway necessary for communication makes use of all the sensory modalities – vision, hearing, smell, taste and touch – and, in a sense, communication represents a final common pathway through which these sensations are co-ordinated and expressed.

Presumably, evolution intended these messages to be direct and unambiguous. There was, no doubt, also the intention to keep them secret within the confines of a particular species. Yet from the outset they have been subjected to espionage and deception as nature perversely subverted her own strategies: a prime example of how nature has always worked, where one adaptation or strategy is succeeded by another, nature always outdoing herself.

Strategies of deception

We have seen how the bolus spider can mimic the sex pheromone of the moth on which it preys (see Chapter 10), but mimicry also extends to structural imitation. In an attempt to avoid predation, the colour patterns on the wings of some species of butterfly correspond to those found on other species which are known to be unpalatable to predators or full of toxins – so-called Batesian mimicry. And the cuckoo lays an egg that is a perfect match to those of the host species that it parasitises.

Eavesdropping is widely practised in the insect world, commonly by predators, which are able to detect and recognise pheromone signals

A low-ranking individual utters an alarm call designed to scatter the troop, thus leaving behind the bulk of the food for the deceiver.

produced by prey species. Bark beetles are small creatures that infest the bark of certain trees in large numbers; having found a suitable tree, the beetle secretes an 'aggregation' pheromone, which attracts others of its kind. But this pheromone is 'eavesdropped' by a number of different predators, which then set about decimating beetle numbers.

Behaviour worthy of a prize for initiative goes to those insects that employ chemical camouflage to gain access to their prey as 'wolves in sheep's clothing'. Certain species of ant enslave honeydew aphids and, by stroking them in a particular fashion ('milking' them), cause the aphids to release a drop of sugar-rich secretion, which is an excellent food source for the ants. The larvae of the lacewing butterfly feed on these aphids but, to harvest this resource, they must pass through the ranks of ants guarding their slaves. To do this, they cover themselves in waxy secretions produced by the aphids themselves. From the point of view of the ants, the butterfly larvae may not look like aphids but, if they smell like aphids, then presumably they must be aphids.

Overall, deception and deceit are uncommon in non-human primates.

They are most frequently observed in the great apes (chimpanzees, gorillas and orangutans), but these findings need to be treated with some caution since they stem largely from studies undertaken with captive animals or in a laboratory setting. Undoubtedly, deception does occur in the wild but is probably far less common than suggested by laboratory data. In his book *Wild Minds*, Marc Hauser quotes his observation of a vervet monkey emitting an alarm call, usually associated with the presence of a leopard, while being pursued by a group of vervets seemingly intending violence upon it. Since alarm calls are always taken seriously, it had the effect of breaking off the attack immediately, thereby allowing the fleeing monkey to avoid pain and punishment. This form of deception, with the use of false alarm signals, has been documented in other circumstances, where a low-ranking individual, with limited access to a supply of food, utters an alarm call designed to scatter the troop, thus leaving behind the bulk of the food for the deceiver. You may think this strategy would only succeed on one or two occasions, but such is the imperative of the alarm call that, in fact, it works time and time again. Silence may also be employed as a tool of deception, especially to conceal a hidden cache of food; a monkey may sit poker-faced, feigning lack of interest or maintaining its gaze on anything other than, for instance, a tasty morsel it has managed to secure.

Outdoing the animals

Sophisticated though human society may be, we are also the outright winners of the prize for dodgy behaviour. Deceit, in varying degrees, has brought about the downfall of individuals and governments, cost millions of lives in conflicts based on illegitimate claims and fostered the rise of totalitarian regimes built on a culture of fear, cynically engineered by falsehoods. As perhaps never before, we find ourselves caught up in a world where fact and fiction often blend seamlessly and truth is a hard commodity to find. Nowhere in the biological and ecological sciences have these issues gained more prominence than in matters concerning the active management of African wildlife. With the future of much of the world's wildlife hanging in the balance, it is worth looking more closely at the example of the African elephant (see Chapter 12), where the waters have been particularly muddied, not just by those with a vested commercial interest, but also by reputable biologists and conservationists.

CHAPTER FOURTEEN

IS ANYONE HOME?
animal thinking

Are individual animals essentially robots driven by instinct in pursuit of equally robotic prey species?

Perceiving the world around us

Sunrise one morning in November found me in the company of the Selinda pride of lions out on the flood plain of the Kwando river in Northern Botswana. Originally 23 in number, the pride had not seen its leaders, a coalition of three males, for some months (the coalition possibly having deserted the pride to take over another one elsewhere). Not deterred in any way by the lack of males, the large core group of females and their offspring enjoyed possession of a substantial territory with easy access to food.

This morning started as usual with most of the lions lounging in the grass soaking up the warmth of the early sun, cubs active and playful, adults relaxed and indulgent. A short distance away a herd of buffalo were grazing on the edge of a reed bed, ignoring the lions and in turn

The lions were gazing intently at two warthogs busily digging in the sandy soil.

being ignored by them. I found myself seduced into a reverie by the sight of prey and predators almost together, so that it took me a little while to realise that the lions had shaken off their lethargy and were gazing intently at two warthogs busily digging in the sandy soil for grass roots and tubers. With no apparent exchange of signals, two of the lionesses moved off in a wide circular movement, taking them to a position behind the warthogs. Once this was done two further groups moved off right and left to flanking positions, and the trap was finally sprung by the remaining lions simply walking out in the open towards the warthogs.

Seeing this drama unfold in real life is a vastly different experience from reading about it or even seeing it on film. Not for the first time, this behaviour raised for me questions about how lions might perceive the world around them. What mental forces are in operation during hunting, where a whole group of animals co-ordinate their activities to secure a successful outcome? How do they arrive at the decisions that shape their daily lives? Are they consciously aware of themselves as individuals or are they essentially robots driven by an instinct to act out the role of predator as they chase and kill equally robotic prey species?

What is consciousness?

It is difficult enough at times to understand the thoughts of other people, including those with whom we enjoy a close personal relationship. In this light, how much more difficult might it be to try and climb inside the mind of an entirely different species? How can we possibly judge what an animal may think of itself or how it sees the world in which it lives? Any attempt to unravel these questions is, to say the least, a formidable undertaking, and it has to be said from the outset that scientific understanding in this regard is incomplete.

In order to discuss this question of consciousness it is necessary to come to some sort of definition of what we are talking about. The word 'conscious' comes from two Latin roots: 'con', meaning with and 'scire', meaning to know. A literal translation of consciousness therefore becomes 'with knowledge'.

Given the context in which we are using the word, it is obvious that a large number of species can be said to be conscious – they have a knowledge of the world around them delivered through the medium of their five senses (sight, sound, taste, hearing and touch). Most biologists would have no quibble with this, but the real problem comes when we start to talk about self consciousness or self awareness. This concept carries knowledge to a higher level altogether, implying that there is not only a knowledge of the external world but also an awareness that the creature knows itself to be an independent, separate entity – that 'I' am 'me' and nobody else, similar to, but distinct from, all others of my kind.

Merely to suggest that animals may not only think, but also act on those thoughts is to invite ridicule from most scientists engaged in the study of animal psychology. To their way of thinking, even the most complex forms of animal behaviour can be reduced to a series of small steps, each governed by instinctive (unconscious) responses laid down in the genetic material of each species. And there are some examples which seem to confirm this viewpoint, including nest-building behaviour in birds and the complex patterns often associated with mating displays.

If any impact is to be made against this seemingly unassailable fortress of scientific opinion, how might it be done? The most obvious approach is to study the evolution of the brain, looking for the first appearance of structures that we know to be important in humans for conscious reflection. A second strategy would be to examine aspects of animal behaviour, trying to identify patterns of action that could only proceed with conscious knowledge.

The three-in-one brain

The belief that self consciousness is the sole prerogative of humans brings with it an awkward question: why should evolution have restricted this faculty to only one species? Natural selection involves the gradual development of a whole raft of characteristics that enable animals to adapt to particular environments. The evolution of the human eye, for example, can be traced back for literally hundreds of millions of years to a time when it started as a simple collection of cells, in primitive organisms, which were capable of responding to light. On this basis it seems highly unlikely that something as complex as consciousness should 'fall from the sky', already neatly packaged for the benefit of a single species.

The evolution of the human brain can be represented as the construction of a three-storey house, each level possessing its own distinct history. The ground floor of such a building corresponds to the oldest and most primitive area of the brain – the common inheritance of all vertebrates from the simplest to the most advanced. Within this ancient nervous system were discrete groups of cells which over aeons of time began to develop and evolve into highly complex structures, allowing some groups – particularly reptiles, birds and mammals – a much more sophisticated appreciation of their environment, together with a greater flexibility in behaviour. These complex cell structures were destined to become most dominant in mammals, especially in primates and humans. They came to represent the upper floors of the evolutionary house.

Consciousness requires an awareness
that 'I' am 'me' and nobody else.

Housed within the ground floor, known as the hind-brain, are specialised centres which are concerned with the automatic control of a variety of body functions including breathing, temperature regulation and cardiovascular reflexes. At this evolutionary level behaviour is essentially instinctive in nature. Actions are driven by compulsive need so that strategies of defence, aggression and territorial acquisition are well developed. Despite the sophistication of the human brain we are not completely immune to the demands of this ancient inheritance, as a glimpse at any newspaper headlines will readily testify.

But as far as the emergence of consciousness is concerned, it is the second story of the house that holds most of the clues. This floor is known as the mid-brain or paleomammalian brain on account of its appearance and original development in the earliest mammalian ancestors (mammals originally evolved from a group of small reptiles, the cynodonts). Similar features emerged in birds during their evolution from another reptilian group. The addition of this further brain capacity, present in all mammals and birds, introduced further developments in perception and behaviour extending beyond the limits imposed by instinctive imperatives.

With this new ability, behaviour that we would classify as emotion became a prominent feature, including pleasure and fear, social bonding and maternal attachment. Also located within this area of the brain are a group of closely connected structures, collectively known as the limbic system, including the oldest representatives of a well defined cerebral cortex. And it is within the workings of the limbic system that self-awareness first became a possibility. The key that was to open the door was the emergence of memory, which allowed behaviour to throw off the shackles of purely reflex action and made possible behaviour that could be based on the knowledge gained by past experience.

The top story of the house, the neo or new mammalian brain, is composed of a very highly developed cerebral cortex, containing an estimated 100 billion nerve cells in humans. Its function is to collate and interpret signals delivered from all areas of the body and to formulate and execute an appropriate response. A network of connections between different areas of the cerebral cortex allows extensive interaction between different areas of the brain, and it is this feature in humans that confers a major advance in our capacity for thought, learning and the performance of complex actions. Although the cerebral hemispheres are present in all vertebrates it is only in birds, mammals and modern reptiles that a high degree of complexity has been achieved.

The anatomy of memory

It is now known that the ability to store memory resides largely in one of the components of the limbic system – the hippocampus, so named by early anatomists because its shape resembles the tail of a seahorse (in Greek, 'hippos' means horse, and 'kampos' means sea monster). Disease of the hippocampus in humans is rare, but if it is damaged or destroyed, the individual loses all ability to store any new memory for more than a few minutes, as well as being afflicted by a total failure to learn new material based on the written or spoken word. These people live a miserable existence, perpetually confronting a new world every few minutes with no sense of any personal past.

According to the modern view, memory is laid down in a series of neural (nerve cell) networks or maps, each containing up to 10 000 individual neurones. And there may be as many as 100 million such groupings in the human brain, each of which contains a specific memory fragment.

The hippocampus has the job of filtering the vast amount of sensory information bombarding the brain every minute of every day, and passing some of this on to other centres, discarding some, but retaining that which is destined to enter memory (including the spoken and written word). What is so remarkable about the hippocampus is its ability to replay loops of memory in a series of feedback circuits until the memory is retained in permanent storage. But what does this have to do with consciousness?

From memory to consciousness

Suppose, for the sake of argument, you had driven your car along a dark, wet road some six months ago, misjudged a corner in the slippery conditions and found yourself driving into a ditch. Driving along that same road tonight in similar conditions you would hopefully slow down to avoid a repetition of the accident. Or, to put it another way, your memory of that first occasion would make you conscious of the dangers you face today. You then make a conscious decision to take steps to avoid a recurrence. Memory therefore is critical in the chain of events leading to consciousness.

Gerald Edelman, a scientist at the Rockefeller University (USA) puts it this way: consciousness involves 'the recognition by a thinking subject of his or her own acts and affectations. It embodies a model of the personal and of the past and future as well as the present.' Edelman termed this ability 'higher order consciousness' and, because it frequently employed the use of language and the awareness of past events, he felt it to be uniquely human. The assumption here, of course, is that other animal species are incapable of employing language and have no memory for past events. As it happens this is incorrect on both counts.

Language, as vocal communication, is in wide use throughout the animal kingdom. This has already been discussed in relation to elephants (see Chapter 11) but it might equally apply to birdsong, the chatter of monkeys and the sounds made by whales and dolphins. Animals as varied as the jackal and the squirrel hide caches of food during times of plenty, only to retrieve them later when food becomes scarce. And the Clark's nutcracker, a bird resident in the American south-west, has been observed to store as many as 33 000 pine nuts in as many as 6 000 different locations during the late summer, returning to reclaim them with remarkable accuracy during the winter. None of this behaviour would be possible without memory. It follows that any species capable of possessing memory is also capable of utilising this information in modifying its behaviour on the basis of past experience – a sequence not possible without conscious appraisal.

Different ways of seeing the world

All animals need to construct a picture of the world in which they live. This is necessary as a basic matter of survival – they need to know what places and which creatures to avoid, as well as recognising likely locations where food and mating opportunities may be found. Animals build up this picture through the information delivered through their sense organs. And because sense organs vary in their relative importance in different species, it follows that each species sees the world in somewhat different terms. The world of the bat is largely appreciated through acoustic images, its ears being the dominant sense organ. Even deprived of sight, a bat can live quite normally, finding its way around and feeding by virtue of echo location alone. The same cannot be said of the fish eagle where the primary sense organ is the eye. And although a fox and a whale are both mammals they can hardly be expected to view the world in the same way. As a consequence, we should expect conscious appreciation of the world to show differences according to species.

Humans are no exception to this and we see the world in our own unique way. Unfortunately, in studying other species we seem to have an irrepressible need to compare their behaviour to a self-styled human gold standard. This is the way we have come to an opinion as regards their 'worth', intelligence, motor skills, etc. And this makes it very difficult to study animal behaviour with real objectivity because we always have the tendency to interpret their actions in terms of human motivation.

But what makes sense to an animal may make no sense to us. In Chapter 13 we discussed the mate selection behaviour of the black grouse: the remarkable ability of the female to predict the future lifespan of its potential suitor. Scientists have no idea how this is achieved. In coming to terms with the prospect of conscious animal intelligence we have to accept that a major problem lies in our own ignorance, our own inability to view the world through non-human eyes.

Through a mirror dimly

At some time during the 1870s, Charles Darwin presented a large mirror to two orangutans resident in the London Zoo. Seeing the reflection of apes in the mirror the orangutans were observed to press and rub the mirror, grimace at it, kiss the surface, arrange themselves in various positions in front of it, look behind it and finally ignore it. As these observations stand it is difficult to know what to make of them. But 100 years later, Gordon Gallop, a comparative psychologist, thought that this set-up might be a way of investigating self awareness in animals. In other words if they were capable of recognising their mirror image then they must be self aware. To this end he painted red dots over one eye and one ear of chimpanzees when they had been rendered unconscious by anaesthesia. Subsequently placed in front of a mirror, the chimpanzees immediately touched the areas of the red dots on their own bodies. Clearly they appeared to recognise their own self image.

Repeated over a spectrum of species, red dot recognition was found to occur in some, but by no means all, chimpanzees, orangutans, bonobos, one single gorilla and children over the age of two years. No other tested species reacted to the presence of red dots. Gallop concluded that self awareness was therefore limited to humans and a few primates only.

On the face of it this sounds plausible enough, but what does it really mean? For a start the possession of mirrors is a purely human affectation. Nowhere in evolutionary biology does the recognition of one's face carry any advantage. Facial self recognition and mirror images are not part of the conscious world constructed by any species other than humans – the only time an animal might see its mirror image is when it stoops to drink in a clear stream. Failure in the mirror test, beyond a certain modicum of curiosity, is more likely to reflect an absence of meaning rather than a lack of self consciousness. And there is a curious parallel here in the human medical condition of prosopagnosia, a disorder of the nervous system in which the individual lacks the capacity to recognise his/her own face in the mirror, yet is well aware of himself/herself as an individual with a clear concept of self.

Animal behaviour and consciousness

Confronted by all these observational difficulties and the minefield presented by attempts to interpret the behaviour of non-human species, is there any way that would allow us once and for all to say that many species enjoy a conscious existence? As it happens, there is, and these patterns of behaviour are to be most easily found among species that form social groups. They include the use of language, flexibility in behaviour, co-operation between individuals and the capacity to deceive.

A question of language

Language is a form of communication used to transmit information between members of a species. Broken down into its component parts, language implies firstly that a thought has been generated and that there is a perceived need to transmit this information to another. Secondly, the transmission itself is required to carry meaning and is capable of being understood and acted upon by the recipient.

The pioneering work of Katherine Payne and Joyce Poole on elephant vocalisation (see Chapter 11) has achieved wide acclaim in the general press, yet has been met by deafening silence in the ivory towers of academic behavioural science. Elephant communication meets all the criteria of language outlined above, but clearly represents a bridge too far for academics. The same criteria apply to birdsong, monkey chatter, etc, and are an indication that the human concept of oral language requires serious revision.

Flexible behaviour

If we accept for a moment that all animal behaviour is unconscious and proceeds through a series of automatic steps without any measure of reflection, then we cannot escape the conclusion that behaviour must be rigid and stereotyped – constant and without variation in any given circumstance. This version of events, however, was turned upside down when it was discovered that a variety of creatures were capable of using tools – a faculty previously thought to be unique to humans because it implies reasoned intent, hardly possible if there is no consciousness of what is being achieved or why the act is being performed.

Chimpanzees 'fish' for termites by stripping leaves off a small branch and poking it into the termite mound. All it takes is a few minutes for a number of termites to adhere to the stick, thus providing a readily available high-protein meal. In similar fashion the woodpecker finch employs a sharp twig to spear insect larvae hidden under the bark of trees. Out on the African plains the Egyptian vulture is one of many birds to employ stones to break open large eggs. And otters always use stones to crack open shellfish, often carrying around a favourite stone for several days.

Otters always use stones to crack open shellfish, often carrying around a favourite stone for several days.

Co-operative ventures

The Selinda lions at the start of this chapter did not engage in a free-for-all chase of the warthogs, but proceeded to lay a trap that required the co-ordinated activity of a number of lions.

A co-operative venture supposes that each member of the group is aware of the intention of others – a failure to do so will result only in chaos. But this is only part of the story, because achieving a successful outcome demands of each lion a knowledge of its own individual role. Although the overall outline of such a planned hunt may owe something to genetic inheritance and imitation, it stretches belief too far to accept that the lions do not consciously know what they are doing or how their individual contribution fits into the scheme of things. Some degree of flexibility of behaviour is required at all times, since their prey does not behave like a robot and tries every means at its disposal to escape, in turn requiring the predators to adapt their approach to any new situation as it arises.

The art of deception

As any con-artist will testify, successful deception needs to meet certain requirements. Firstly, behaviour that is likely to deceive requires to be plausible, to conform to an accepted normal repertoire. Bizarre or unusual behaviour may arouse curiosity but is unlikely to deceive. Secondly, deceit must be used sparingly because frequent use will simply expose it for the lie that it is. And, finally, the purpose of deceit must of course be to gain advantage.

The eggs and young of ground-nesting birds such as plovers and sandpipers are highly vulnerable to a number of predators, including weasels, stoats and foxes. In an attempt to distract predators that wander too close to the nest, one of the parents frequently feigns injury. Fluttering a few metres away, it will drag its tail or wing along the ground, uttering distress calls. This sequence may continue until the predator has been lured away from the danger zone. The strategy is used only when the nest contains eggs or young, for at other times the bird will simply fly away. An alternative approach is to engage in 'false incubation', where one of the parent birds sits on the ground at a safe distance from the nest when a predator approaches. This may be a sufficient distraction but if it is not then the bird will resort to feigned injury.

When threatened by predators, vervet monkeys produce an alarm call that varies according to the type of predator. Such is the imperative of an alarm call that it always invokes a flight response. It can therefore be used as a basis for deception to allow, for example, a sneaky individual better access to a communal food source or to avoid punishment from a dominant individual.

A meaningful existence

Consciousness likely arose when the individual components of the limbic system came together during evolution to function as a co-ordinated whole.

Since this system is present in all birds and mammals, conscious awareness should be within the compass of all of their members. Whether some measure of consciousness applies to simpler creatures is not known, simply because the workings of their nervous systems is poorly understood. What makes consciousness in humans so special is the tremendous elaboration of the cerebral cortex, making the range of thought and behaviour so much wider than in other mammals. Which is not to say that other mammals should be relegated to some second division.

Animals form a picture of the world appropriate to their sensory perceptions and way of life. It is surely not difficult to accept that any thoughts they may have are likely to be of a different quality to those of humans, who view the world in an entirely different way. An insistence that the construction and elaboration of thought processes must proceed along human lines is to believe that all automobiles are powered by the same engine.

One of the unfortunate side effects of human consciousness has been our dissociation from the world of nature, fostering a belief that we are a superior form of life. One step along the way to puncturing our balloon of hubris might be a realisation that species other than our own may enjoy a conscious, 'meaningful' existence.

Otherwise we may be consigned to the role of mere consumers rather than active participants in a drama where the business of survival and adaptation are acted out on many stages, where no one form of life possesses any greater intrinsic value than another.

BIBLIOGRAPHY

Over 250 scientific papers and books were consulted in writing this book; the
bibliography shown is a representation of the wide scope of the relevant literature.

Alvarez W, Asaro F, 1990. 'An extraterrestrial impact'. *Scientific American* 263: 44–52.

Bekoff M, 2002. *Minding animals: Awareness, emotion and heart*. Oxford University Press, UK.

Bell R, 1971. 'A grazing ecosystem in the Serengeti'. *Scientific American* 224: 86–93.

Bull M, Kennedy – Stoskopf S, Levine J *et al*, 2003. 'Evaluation of T lymphcytes in captive African lions (Panthera leo) infected with feline immunodeficiency virus'. *American Journal of Veterinary Research* 64: 1293–1300

Darwin C, 1859. *On the origin of species by means of natural selection*. John Murray, London.

Dawkins M, 1993. *Through our eyes only? The search for animal consciousness*. WH Freeman/Spektrum, Oxford, UK.

Eldredge N, Gould S, 1972. 'Punctuated Equilibrium: an alternative to phyletic gradualism'. *Models in paleontology* (Schopf T, editor). Freeman, Cooper & Co, San Francisco, USA.

Essex M, Kanki P, 1988. 'The origin of the Aids virus'. *Scientific American* 259: 64–71.

Fayrer-Hosken R, Grobler D, van Altena J *et al*, 2000. 'Immunocontraception of African elephants'. *Nature* 407: 149

Frank L, Davidson J, Smith E, 1985. 'Androgen levels in the spotted hyena'. *Journal of the Zoological Society of London* 206: 521–531.

Goetz R, Keen E, 1957. 'Some aspects of the cardiovascular system in the giraffe'. *Angiology* 8: 542–564.

Grant P, 1991. 'Natural selection and Darwin's finches'. *Scientific American* 265: 60–65.

Griffin D, 1984. *Animal Thinking*. Harvard University Press, Cambridge, USA.

Harvey C, Kat P, 2000. *Prides: The lions of Moremi*. Southern Book Publishers, Rivonia, South Africa.

Hofmann-Lehmann R, Fehr D, Grob M *et al*, 1996. 'Prevalence of antibodies to feline parvovirus, calicivirus, herpesvirus, coronavirus and immunodeficiency virus and of feline leukaemia virus antigen and the interrelationship of these viral infections in free-ranging lions in East Africa'. *Clinical and Diagnostic Laboratory Immunology* 3: 554–582.

Janis C, 1976. 'The evolutionary strategy of the edquidae and the origins of rumen and cecal digestion'. *Evolution* 30: 757–774.

King A, McLelland J, 1984. *Birds: Their structure and function*. Baillière Tindall, London, UK.

Kyle W, 1992. 'Simian retroviruses, poliovaccine and origin of AIDS'. *Lancet* 339: 600–601.

MacLean P, 1973. *A triune concept of the brain and behaviour*. University of Toronto Press, Canada.

Neaves W, Griffin J, Wilson J, 1980. 'Sexual dimorphism in the phallus of the spotted hyena'. *Journal of Reproduction and Fertility* 59: 509–513.

Officer C, Drake C, 1985. 'Terminal Cretaceous environmental events'. *Science* 227: 1161–1167.

Olmstead R, Langley R, Roelke M *et al*, 1992. 'Worldwide prevalence of lentivirus infection in wild feline species: Epidemiologic and phylogenetic aspects'. *Journal of Virology* 66: 6008–6018.

Packer C, Altizer S, Appel M *et al*, 1999. 'Viruses of the Serengeti: patterns of infection and mortality in African lions'. *Journal of Animal Ecology* 68: 11261–1178.

Payne K, 1989. Elephant talk. *National Geographic* (August issue) 266–276.

Poole J, 1996. *Coming of age with elephants*. Hodder & Stoughton, London, UK.

Raup D, 1988. 'Extinctions in the geological past'. *Origins and Extinctions* (Osterbrock D & Raven P, editors). Yale University Press, New Haven, USA.

Roelke M, Pecon-Slattery J, Taylor S *et al*, 2006. 'T-Lymphocyte profiles in FIV-infected wild lions and pumas reveal CD4 depletion'. *Journal of Wildlife Diseases* 42: 234–248.

Stanley S, 1987. *Extinction*. Scientific American Library, New York, USA.

Stern K, McClintock M, 1998. 'Regulation of ovulation by human pheromones'. *Nature* 392: 177–179.

Taylor C, 1970. 'Dehydration and heat: effects on temperature regulation of East African ungulates'. *American Journal of Physiology* 219: 1136–1139.

Van Citters R, Kemper W, Franklin D, 1966. 'Blood pressure responses of wild giraffe studied by radiotelemetry'. *Science* 152: 384–386.

Van Lawick-Goodall J, 1970. 'Tool use in primates and other vertebrates'. *Advances in the study of Behaviour*, Vol 3 (Lehrman D, Hinde R, Shaw E, editors). Academic Press, New York, USA.

Von Frisch K, 1967. *The dance language and orientation of bees*. Oxford University Press, UK.

Willis K, McElwain J, 2002. *The evolution of plants*. Oxford University Press, UK.

Wyatt T, 2003. *Pheromones and animal behaviour*. Cambridge University Press, UK.

INDEX